中国高等教育学会2022年度高等教育科学研究规划课题“新农科普通高等学校涉农专业理科基础课程体系、课程思政与教材建设研究(22NL0418)”项目教材

Quantitative Analytical Chemistry

Editor-in-Chief:
Li Xingyu
Qin Xiangdong

中国农业大学出版社
China Agricultural University Press
· Beijing ·

内 容 简 介

Quantitative analytical chemistry is a cornerstone course in the undergraduate curriculum at higher agricultural institutions. The comprehensive coverage of this textbook encompasses fundamental knowledge, classical methods, and modern techniques in quantitative analytical chemistry. Each analytical method is meticulously explained, with detailed discussions on principles, operational steps, data processing, and result interpretation. The textbook includes numerous practical examples and exercises to reinforce and apply the acquired knowledge, emphasizing theoretical exposition, and practical application. This textbook is an ideal bilingual teaching resource for undergraduates studying biotechnology, agronomy, plant protection, food science, environmental science, animal science, horticulture, tobacco science, applied chemistry, and bioengineering at higher agricultural and forestry institutions. Additionally, it is an invaluable reference for graduate students and scientific researchers specializing in chemical analysis.

图书在版编目(CIP)数据

定量分析化学＝Quantitative Analytical Chemistry：英文 / 李兴玉，秦向东主编．-- 北京：中国农业大学出版社，2025. 5. -- ISBN 978-7-5655-3394-5

Ⅰ．O655

中国国家版本馆 CIP 数据核字第 2025N60S82 号

书　　名　Quantitative Analytical Chemistry
作　　者　Li Xingyu　Qin Xiangdong　主编

策划编辑　何美文　赵　艳　　　**责任编辑**　赵　艳
封面设计　中通世奥图文设计中心
出版发行　中国农业大学出版社
社　　址　北京市海淀区圆明园西路 2 号　　　**邮政编码**　100193
电　　话　发行部 010-62733489,1190　　　读者服务部 010-62732336
编辑部 010-62732617,2618　　　出　版　部 010-62733440
网　　址　http://www. caupress. cn　　　**E-mail** cbsszs@cau. edu. cn
经　　销　新华书店
印　　刷　河北虎彩印刷有限公司
版　　次　2025 年 5 月第 1 版　　2025 年 5 月第 1 次印刷
规　　格　185 mm×260 mm　　16 开本　　15 印张　　477 千字
定　　价　48.00 元

Editorial Committee

Editor-in-Chief:

Li Xingyu 李兴玉 (Yunnan Agricultural University)

Qin Xiangdong 秦向东 (Yunnan Agricultural University)

Associate editors:

Zhao Lei 赵　蕾 (Yunnan Agricultural University)

Shen Xiaojing 沈晓静 (Yunnan Agricultural University)

Huang Lulu 黄璐璐 (Yunnan Agricultural University)

Huang Hongping 黄红苹 (Yunnan Agricultural University)

Li Song 李　崧 (Yunnan Agricultural University)

Shahzad Munir (Yunnan Agricultural University)

Editorial members:

Deng Linxin 邓临新 (Yunnan Police College)

Jiang Weiwei 姜薇薇 (Yunnan Agricultural University)

Ma Junrong 马俊蓉 (Yunnan Agricultural University)

Wu Yamei 吴娅梅 (Yunnan Agricultural University)

Yuan Wenjuan 袁文娟 (Yunnan Agricultural University)

Zhao Hongmei 赵红梅 (Yunnan Agricultural University)

Zhao Yan 赵　艳 (Yunnan Agricultural University)

Reviewer:

Yang Xiaodong 羊晓东 (Yunnan University)

Preface

In the ever-evolving field of chemistry, analytical chemistry stands as a cornerstone, bridging the gap between theoretical knowledge and practical application. With great enthusiasm and a profound sense of responsibility, we present *Quantitative Analytical Chemistry*, a textbook designed to illuminate the path for students embarking on the journey of understanding the intricate world of analytical chemistry. This textbook culminates years of collaborative efforts, insights, and expertise from a dedicated team of authors at Yunnan Agricultural University. Our collective vision has been to construct a resource that educates and inspires the next generation of interdisciplinary chemists and scientists.

The genesis of this textbook was motivated by a clear need for an introductory yet comprehensive guide to analytical chemistry that caters to the requirements of chemistry majors and related fields, such as life sciences, environmental sciences, and agriculture. Recognizing the diverse applications and the critical role of analytical chemistry in these fields, we aimed to create a textbook that balances theoretical principles with practical applications, thereby equipping students with the knowledge and skills necessary to navigate and contribute effectively.

Guided by Xi Jinping Thought on Socialism with Chinese Characteristics for a New Era, this textbook also reflects the spirit of the 20th National Congress of the Communist Party of China, which emphasizes the importance of education in building a strong nation and cultivating high-quality talent. In line with the Congress's call to deepen comprehensive reforms in education, this textbook integrates innovative teaching methods and cutting-edge research to meet the strategic needs of China's modernization and agricultural development. By incorporating interdisciplinary approaches and fostering critical thinking, the textbook aims to cultivate students' ability to solve real-world problems, contribute to scientific and technological self-reliance, and support the goals of rural revitalization and green development.

Our primary objective has been to lay a solid foundation in the principles of analytical chemistry, ensuring that students gain a robust physical understanding of the subject. Deep comprehension of these principles is paramount for students to apply their knowledge effectively in real-world scenarios. Furthermore, we have endeavored to cultivate an appreciation for the complexities in evaluating experimental data's accuracy and precision. Through this textbook, we introduce statistical methods as essential tools for refining these evaluations, thereby enhancing the reliability of analytical results.

In addition to foundational knowledge, this textbook aims to familiarize students with a spectrum of analytical techniques, both modern and classic. We aim to train students to analyze these methods and understand their underlying principles and applications. By doing so, we hope to foster versatile analytical chemists who can adapt to and innovate within the field.

The structure of *Quantitative Analytical Chemistry* reflects our holistic approach to teaching analytical chemistry. This textbook is divided into 10 chapters. Firstly, it introduces the subject and then delves into error and data analysis, titration analysis, gravimetric analysis, photometry, and potentiometric analysis. Each chapter aims to start with clear learning objectives, guide students to complete materials, focus on key points, and optimize the learning experience.

The textbook exemplifies remarkable collaborative efforts, underscoring its unique strength. Under the editorial guidance of Li Xingyu and Qin Xiangdong, it brings together a diverse range of insights and expertise thanks to the contributions of a distinguished team of authors. The textbook's composition showcases this collaborative spirit, with chapters 1, 2, 3, and 7 jointly completed by Li Xingyu, Huang Hongping, Qin Xiangdong, and Ma Junrong. Chapters 4 and 9 embodied Li Song, Deng Linxin, and Shen Xiaojing's contributions, while Zhao Hongmei and Zhao Yan co-authored chapters 5 and 9. Huang Lulu and Zhao Lei wrote chapter 6, chapter 8 and the appendix. Chapter 10 was a joint effort by Wu Yamei, Jiang Weiwei, and Yuan Wenjuan. Overseeing the coherence and quality of the entire work, Shahzad Munir played a pivotal role in reviewing all the content, ensuring the textbook's integrity and value to its readers.

The publication of our textbook went smoothly, and we are grateful for the support and contributions of many individuals and organizations. Firstly, we would like to express our heartfelt gratitude to Professor Yang Xiaodong of Yunnan University for his thorough reading and comprehensive textbook review. Secondly, we thank Ms. He Meiwen and Ms Zhao Yan, the editors of China Agricultural University Press, for their meticulous planning and editing. Additionally, we appreciate all the members of the Editorial Committee for their hard work and dedication. Our textbook was generously supported by several projects and institutions, including the Yunnan Agricultural University Undergraduate Education and Teaching Reform Research Project (YNAUJG2023031), the Agricultural Chemistry teaching team of Yunnan Agricultural University's Huang Danian-style teaching team, the College of Science of Yunnan Agricultural University, and China Agricultural University Press. We hereby express our profound gratitude to all of them for their invaluable assistance.

As educators and authors, we are keenly aware of the importance of continuous improvement and adaptation in educational resources. Therefore, we warmly welcome comments and suggestions from our readers. Your insights are crucial for guiding this textbook's future editions, enabling us to better serve the needs of students and educators.

In closing, we sincerely hope that *Quantitative Analytical Chemistry* serves as a comprehensive guide for students and a source of inspiration. May this textbook ignite a passion for analytical chemistry and empower students to contribute meaningfully to the scientific community. Welcome to the fascinating world of analytical chemistry, where every analysis promises discovery and advancement.

Li Xingyu, Qin Xiangdong, and the Authorial Team
August 2024

Contents

Chapter 1

Introduction

Key Points

- *Tasks and Functions of Analytical Chemistry*
- *Classification of Analytical Chemistry Methods*

1.1 Tasks and Functions of Analytical Chemistry

Analytical chemistry studies and uses instruments and methods to separate, identify, and quantify matter. In practice, separation, identification, or quantification may constitute the fundamental analysis or be combined with another method. Separation isolates analytes. Qualitative analysis identifies analytes, while quantitative analysis determines the numerical amount or concentration.

Analytical chemistry consists of classical, wet chemical, and modern instrumental methods. Classical qualitative methods use separations such as precipitation, extraction, and distillation. Identification may be based on color, odor, melting point, boiling point, radioactivity, or reactivity differences. Classical quantitative analysis uses mass or volume changes to quantify the amount. Instrumental methods may be used to separate samples using chromatography, electrophoresis, or field flow fractionation. Then qualitative and quantitative analysis can be performed, often with the same instrument, and may use light, heat, electric, or magnetic fields. The same instrument can often separate, identify, and quantify an analyte.

Analytical chemistry also focuses on improvements in experimental design, chemometrics, and the development of new measurement tools. Analytical chemistry has broad applications in forensics, medicine, science, and engineering.

The solution to such problems lies within the province of **qualitative analysis** and is outside the scope of the present volume. After determining the properties of a given sample component, analysts are often required to determine the content of each component or specific component. This type of measurement belongs to the field of quantitative analysis, and various techniques can be used to provide the necessary information.

So, analytical chemistry includes qualitative analysis, quantitative analysis, and chemical structure analysis.

1.2 Classification of Analytical Chemistry Methods

(1) *It is divided into qualitative, quantitative, and structural analysis according to the difference in the analysis task.*

(2) *It is divided into inorganic and organic analyses according to the different objects of study.*

(3) *It is divided into constant, semi-trace, trace, and ultramicro analysis according to the quantity of the analytical pattern.*

A major constituent accounts for 1%-100% of the sample under investigation; a minor constituent is present in the range of 0.01%-1%; a trace constituent is one at a concentration of less than 0.01%. With the development of increasingly sophisticated analytical techniques, it has become possible to determine substances present in quantities much lower than the 0.01% upper-level set for trace constituents. It is, therefore, necessary to make further subdivisions: **trace** corresponds to 10^2-10^4 $\mu g \cdot g^{-1}$, **microtrace** to 10^{-1}-10^2 $pg \cdot g^{-1}$, **nanotrace** to 10^{-1}-10^2 $fg \cdot g^{-1}$.

When the sample weight is small (0.1-1.0 mg), determining a trace component at the 0.01% level may be called **subtrace analysis.** If the trace component is at the microtrace level, the analysis is termed **submicrotrace.** With a still smaller sample (not larger than 0.1 mg), determining a component at the trace level is referred to as **ultratrace analysis.** In contrast, with a component at the microtrace level, the analysis is referred to as **ultra-microtrace.**

(4) *It is divided into chemical and instrument analysis according to the analysis principle and the different instruments applied.*

A. **Chemical analysis methods**

Although modern analytical chemistry is dominated by sophisticated instrumentation, the roots of analytical chemistry and some of the principles used in modern instruments are from traditional techniques, many of which are still used. These techniques also form the backbone of most undergraduate analytical chemistry educational labs.

Qualitative analysis: Qualitative analysis determines the presence or absence of a particular compound but not the mass or concentration. Qualitative analyses do not measure quantity.

Chemical tests: Chemical tests are qualitative or quantitative procedures designed to prove the presence or quantify a chemical compound or group using a specific reagent.

Flame test: Inorganic qualitative analysis generally refers to a systematic scheme to confirm the presence of specific, usually aqueous, ions or elements by performing a series of reactions that eliminate a range of possibilities and then confirm suspected ions with a confirming test. Sometimes, small carbon-containing ions are included in such schemes. With modern instrumentation, these tests are rarely used. Still, they can be useful for educational purposes and in fieldwork or other situations where access to state-of-the-art instruments is not available or expedient.

Quantitative analysis: Quantitative analysis measures the quantities of chemical constituents present in a substance.

Gravimetric analysis: Gravimetric analysis involves determining the material present by weighing the sample before and after some transformation. A common example used in undergraduate education is determining the amount of water in a hydrate by heating the sample to remove the water such that the difference in weight is due to the loss of water.

Volumetric analysis: Titration involves the addition of a reactant to a solution being analyzed until some equivalence point is reached. Often, the amount of material in the analyzed solution may be determined. The acid-base titration involving a color-changing indicator is most familiar to those who have taken chemistry during secondary education. However, there are many other types of titrations, such as potentiometric titrations. These titrations may use different types of indicators to reach some equivalence point.

B. **Instrumental analysis methods**

Spectroscopy: Spectroscopy measures the interaction of the molecules with electromagnetic radiation. Spectroscopy consists of many applications, such as atomic absorption spectroscopy, atomic emission spectroscopy, ultraviolet-visible spectroscopy, X-ray fluorescence spectroscopy, infrared spectroscopy, Raman spectroscopy, dual-polarization interferometry, nuclear magnetic resonance spectroscopy, photoemission spectroscopy, Mössbauer spectroscopy and so on.

Mass spectrometry: Mass spectrometry measures the mass-to-charge ratio of molecules using electric and magnetic fields. There are several ionization methods: electron impact, chemical ionization, electrospray, fast atom bombardment, matrix-assisted laser desorption ionization, and others. Also, mass spectrometry is categorized by approaches of mass analyzers: magnetic-sector, quadrupole mass analyzer, quadrupole ion trap, time-of-flight, Fourier transform ion cyclotron resonance, and so on.

Electroanalytical methods: Electroanalytical methods measure the potential (volts) and current (amps) in an electrochemical cell containing the analyte. These methods can be categorized according to which aspects of the cell are controlled and measured. The four main categories are potentiometry (the difference in electrode potentials is measured), coulometry (the transferred charge is measured over time), amperometry (the cell's current is measured over time), and voltammetry (the cell's current is measured while actively altering the cell's potential).

Thermal analysis: Calorimetry and thermogravimetric analysis measure the interaction of material and heat.

Separation: Separation processes are used to decrease the complexity of material mixtures. Chromatography, electrophoresis, and field flow fractionation are representative of this field.

Hybrid techniques: Combinations of the above methods produce a "hybrid" or "hyphenated" technique. Several examples are popular today, and new hybrid techniques are under development. Examples are gas chromatography-mass spectrometry, gas chromatography-infrared spectroscopy, liquid chromatography-mass spectrometry, liquid chromatography-NMR spectroscopy, liquid chromatography-infrared spectroscopy, and capillary electrophoresis-mass spectrometry. Hyphenated separation techniques combine two (or more) techniques to detect and separate chemicals from solutions. Most often, the other technique is some form of chromatography. Hyphenated techniques are widely used in chemistry and biochemistry. A slash is sometimes used instead of a hyphen, especially if the name of one of the methods contains a hyphen itself.

Microscopy: Visualizing single molecules, single cells, biological tissues, and nanomaterials is an important and attractive approach in analytical science. Hybridization with other traditional analytical tools is revolutionizing analytical science. Microscopy can be categorized into three fields: optical, electron, and scanning probe. Recently, this field has been rapidly progressing because of the rapid development of the computer and camera industries.

Lab-on-a-chip: Devices that integrate (multiple) laboratory functions on a single chip of only millimeters to a few square centimeters in size can handle minimal fluid volumes down to less than picoliters.

(5) *It is divided into routine analysis and arbitration analysis according to the different purposes of analysis.*

1.3 General Procedure for Quantitative Analysis

The general procedures of quantitative analysis include the process of sampling, sample pretreatment, determination, calculation, and evaluation of the analysis results.

The results obtained for the proportion of a certain constituent in a given sample may form the basis for assessing the value of a large consignment of the commodity from which the sample was drawn. In such cases, it is essential to be certain that the sample used for analysis represents the whole. When dealing with a homogeneous liquid, sampling presents a few problems. Still, if the material under consideration is a solid mixture, it is necessary to combine some portions to ensure that a representative sample is accurate. The analyst must, therefore, be acquainted with the standard sampling procedures employed for different materials.

With an appropriate sample available, attention must be given to the most suitable technique or techniques for the required determinations. One of the major decisions to be made by an analyst is the choice of the most effective procedure for a given analysis, and to arrive at the correct decision, not only must he be familiar with the practical details of the various techniques and of the theoretical principles upon which they are based, but he must also be conversant with the conditions under which each method is reliable, aware of possible interferences which may arise, and capable of devising means of circumventing such problems. He will also be concerned with questions regarding the accuracy and precision expected from given methods. In addition, he must not overlook the factors such as time and cost. The most accurate method for a certain determination may be lengthy or involve expensive reagents. In the interests of the economy, it may be necessary to choose a method that, although somewhat less exact, yields results of sufficient accuracy in a reasonable time.

Important factors that must be considered when selecting an appropriate method of analysis include: ① the nature of the information that is sought, ② the size of the sample available and the proportion of the constituent to be determined, and ③ the purpose for which the analytical data are required.

The nature of the information sought may require very detailed data, or results of a general character may suffice. Concerning the information which is furnished, different types of chemical analysis may be classified as follows:

(1) Proximate analysis, in which the amount of each element in the sample is determined with no concern as to the actual compounds present.

(2) Partial analysis, which deals with the determination of selected constituents in the sample.

(3) Trace constituent analysis, a specialized instance of partial analysis in which we are concerned with the determination of specified components present in very minute quantity.

(4) Complete analysis when the proportion of each sample component is determined.

Key Glossary

Qualitative analysis: Determines the presence or absence of a particular compound but not the mass or concentration. By definition, qualitative analyses do not measure quantity.

Chemical tests: Chemical tests are qualitative or quantitative procedures designed to prove the existence of or quantify a chemical compound or group using a specific reagent.

Flame test: Inorganic qualitative analysis generally refers to a systematic scheme to confirm the presence of certain, usually aqueous, ions or elements by performing a series of reactions that eliminate ranges of possibilities and then confirm suspected ions with a confirming test.

Quantitative analysis: The quantitative analysis measures the quantities of chemical constituents present in a substance.

Gravimetric analysis: The gravimetric analysis involves determining the material present by weighing the sample before and after some transformation.

Volumetric analysis: Titration involves the addition of a reactant to a solution being analyzed until some equivalence point is reached.

Spectroscopy： Spectroscopy measures the interaction of molecules with electromagnetic radiation.

Mass spectrometry： Mass spectrometry measures the mass-to-charge ratio of molecules using electric and magnetic fields.

Electroanalytical methods： Electroanalytical methods measure the potential (volts) and current (amps) in an electrochemical cell containing the analyte.

Thermal analysis： Calorimetry and thermogravimetric analysis measure the interaction between material and heat.

Separation： Separation processes are used to decrease the complexity of material mixtures. Chromatography， electrophoresis， and field flow fractionation are representative of this field.

Hybrid techniques： Combining the above techniques produces a "hybrid" or "hyphenated" technique.

Microscopy： Visualizing single molecules， single cells， biological tissues， and nanomaterials is an important and attractive approach in analytical science.

Lab-on-a-chip： Devices that integrate (multiple) laboratory functions on a single chip of only millimeters to a few square centimeters in size and can handle minimal fluid volumes down to less than picoliters.

Essential Exercises

Review Questions

1. Briefly explain the tasks and functions of analytical chemistry.
2. How are analysis methods classified?
3. General principles for discussing the selection of analysis methods.

Chapter 2

Experimental Error and Statistics

Key Points

- *Significant Figures*
- *Experimental Error*
- *Statistics*
- *Confidence Intervals*
- *Comparison of Means with Student's Test*
- *Comparison of Standard Deviations with the F test*
- *Test for an Outlier*

Some laboratory *errors* are more obvious, but every measurement error is associated. There is no way to measure the "true" value of anything. The best we can do in chemical analysis is to apply a technique that experience tells us is reliable carefully. Repetition of one measurement method several times tells us the precision (reproducibility) of the measurement. If the result of measuring the same quantity in different ways agrees with one another, then we become confident that the results are accurate, which means they are near the "*true*" value.

2.1 Significant Figures

Suppose that you determine the density of a mineral by measuring its mass (7.235 4 ± 0.000 2) g and volume (2.37 ± 0.02) mL. Density is mass per unit volume: 7.235 4 g/2.37 mL = 3.052 9 g · mL^{-1}. The measured mass and volume are ± 0.000 2 g and ± 0.02 mL, but what is the uncertainty in the computed density? And how many *significant figures* should be used for the density?

2.1.1 Significant Figures in Mass

The number of significant figures is the minimum number of digitsrequired to write a given value in scientific notation without losing precision. The number 158.7 has four significant figures because it can be written as 1.587×10^2. If you write 1.5870×10^2, you imply that you know the value of the digit after 7, which is not the case for 158.7. The number 1.5870×10^2 has 5 significant figures.

The number 5.406×10^{-6} has four significant figures because all four digits are necessary. You could write the same number as 0.000 005 406, which has just four significant figures. The zeros to the left of the five are merely holding decimal places. The number 72 500 is ambiguous. It could mean any of the following: 7.25×10^{5} (3 significant figures), 7.250×10^{5} (4 significant figures), $7.250\,0 \times 10^{5}$ (5 significant figures).

You should write one of the three numbers above, instead of 72 500, to indicate how many figures are known.

Zeros are significant when they occur (5.406×10^{-6}) *in the middle of a number or* (7.250×10^{5}) *at the end of the number on the right-hand side of a decimal point.*

The measured quantity's last significant digit (farthest to the right) always has some associated uncertainty. The minimum uncertainty is ±1 in the last digit. For example, the scale of a Spectronic 20 spectrophotometer is drawn in Figure 2.1. The needle in the figure appears to be at an absorbance of 0.234. We say this number has three significant figures because the numbers 2 and 3 are completely certain, and the number 4 is an estimate. Other people might read the value of 0.233 or 0.235. The percent transmittance is near 58.3 because the transmittance scale is smaller than the absorbance scale at this point. There is more uncertainty in the last digit of transmittance. A reasonable estimate of uncertainty might be (58.3±0.2). There are three significant figures in the number 58.3.

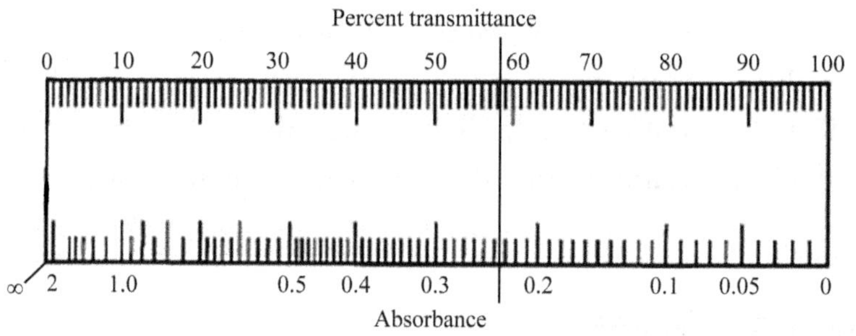

Figure 2.1 Analog scale of Bausch and percent transmittance Lomb Spectronic 20 spectrophotometer. Percent transmittance is a linear scale, and absorbance is a logarithmic scale

When reading the scale of any apparatus, try to estimate to the nearest tenth of a division. For example, on a 50-mL burette, which is graduated to 0.1 mL, read the level to the nearest 0.01 mL. For a ruler-calibrated millimeter, estimate distances to the nearest 0.1 mm.

There is uncertainty in any measured quantity, even if the measuring instrument has a digital readout that does not fluctuate. When a digital pH meter indicates a pH of 3.51, there is uncertainty in digit 1 (and maybe even in digit 5). By contrast, integers are exact. To calculate the average height of four people, you would divide the sum of measurements (a measured quantity with some uncertainty) by the integer 4. There are exactly four people, not (4.000 ± 0.002) people.

2.1.2 Significant Figures in Arithmetic

We now consider how many digits to retain in the answer after you have performed arithmetic operations on your data. Rounding should only be done on the final solution (not intermediated results) to avoid accumulating round-off errors.

1. Addition and Subtraction

If the number to be added or subtracted has equal numbers of digits, the answer goes to the same decimal place as in any of the individual numbers:

$$5.268\times10^{-5}+2.423\times10^{-5}=7.691\times10^{-5}$$

The number of significant figures in the answer may exceed or be less than those in the original data.

$$5.268\times10^{-5}-4.423\times10^{-5}=0.845\times10^{-5}$$

If the numbers being added do not have the same number of significant figures, we are limited by the least certain one. For example, the molecular mass of KrF_2 is known only to the third decimal place because we know the atomic mass of Kr to only three decimal places:

$$18.998\,403\,2(\text{F})\times2+83.798(\text{Kr})=121.794\,\underbrace{8064}_{\text{no significant}}$$

121.794 806 4 should be rounded to as 121.795 as the final answer.

When rounding off, look at all the digits beyond the last place desired. In the preceding example, the digits 804 and 6 lie beyond the last significant decimal place. *Because this number is more than halfway to the next higher digit, we round the 4 up to 5* (that is, we round up to 121.795 instead of down to 121.794). *If the significant figures were less than halfway, we would round down.* For example, 121.794 3 is rounded to 121.794.

If the number is exactly halfway, round to the nearest even digit. Thus, 32.75 is rounded to 32.8 if only three significant numbers exist. If we are retaining only three figures, 4.425×10^{-5} becomes 4.42×10^{-5}. The number $4.425\,001\times10^{-5}$ would become 4.43×10^{-5}, because 5001 is more than halfway to the next digit. *The rationale for rounding to an integer is to avoid systematically increasing or decreasing results through successive round-off errors. Half the round-offs will be up and half down.*

In the addition or subtraction of numbers expressed in scientific notation, all numbers should first be expressed with the same exponent.

$$1.632\times10^{5}+4.107\times10^{3}+9.84\times10^{5}=11.51\times10^{5}$$

The sum $11.513\,07\times10^{5}$ is rounded to 11.51×10^{5} because 9.84×10^{5} limits us to two decimal places when all numbers are expressed as multiples of 10^5.

2. Multiplication and Division

In multiplication and division, *we are generally limited to the number of digits contained in the number with the fewest significant figures*:

$$5.26 \times 10^{-5} \times 2.42 = 1.27 \times 10^{-4}$$

$$6.782\,45 \times 10^{-15} \times 3.8 \times 10^{12} = 2.6 \times 10^{-2}$$

$$72.52 \div 5.168\,47 = 14.03$$

The power of 10 does not influence the number of figures that should be retained. The section on the rule for significant figures explains why it is reasonable to keep an extra digit when the first digit of the answer is 1. The middle product above could be expressed as 2.58×10^{-5} instead of 2.6×10^{-5} to avoid throwing away some of the precision of factor 3.8 in the multiplication.

3. Logarithms and Antilogarithms

If $n = 10^a$, then we say that a is the base 10 logarithm of n:

Logarithm of n: $n = 10^a$ means that $\log_{10} n = a$ **(Equation 2.1)**

Addition and subtraction: express all numbers with the same exponent and align all numbers to the decimal point. Round off the answer according to the number of decimal places in the number with the fewest decimal places. In Equation 2.1, the number n is said to be the antilogarithm of a. The antilogarithm of 2 is 100 because $10^2 = 100$, and the antilogarithm of -3 is 0.001. Your calculator has either a 10^x key or an antilog key. To find the antilogarithm of a number, enter it in your calculator and press 10^x (or antilog).

A logarithm is composed of a characteristic and a mantissa. The characteristic is the integer part, and the mantissa is the decimal part:

$\log_{10} 339 = 2.530$		$\log_{10} 3.39 \times 10^5 = 5.530$	
characteristic	mantissa	characteristic	mantissa
$=2$	$=0.530$	$=5$	$=0.530$

The number 339 can be written as 3.39×10^2. The number of digits in the mantissa of $\log_{10} 339$ should equal the number of significant figures in 339. The logarithm of 339 is correctly expressed as 2.530. The characteristic, 2, corresponds to the exponent in 3.39×10^2. To see that the third decimal place is the last significant place, consider the following results:

$$10^{2.529} = 338\ (338.1)$$

$$10^{2.530} = 339\ (338.8)$$

$$10^{2.531} = 340\ (339.6)$$

The numbers in parentheses are the results before rounding to three figures. Changing the exponent in the third decimal place changes the answer in the third place of 339.

In converting a logarithm into its antilogarithm, the number of significant figures in the antilogarithm should equal the number of digits in the mantissa. Thus,

$$\text{antilog}\ (-3.42) = 10^{-3.42} = 3.8 \times 10^{-4}$$

2 digits 2 digits 2 digits

Here are several examples showing the proper use of significant figures:

$\log_{10} 3.2 = 0.51$	$10^{-2.600} = 2.51 \times 10^{-3}$
$\log_{10} 1\,237 = 3.092\,4$	$10^{4.37} = 2.3 \times 10^{4}$
$\log_{10} 0.001\,237 = -2.907\,6$	$\text{antilog}_{10}\, 4.37 = 2.3 \times 10^{4}$

2.2 Types of Error

Every measurement has some uncertainty, which is called experimental error. Conclusions can be expressed with a high or a low degree of confidence but never with complete certainty. *Experimental error is classified as either systematic or random.*

2.2.1 Systematic Error

Systematic error, also called determinate error, arises from a flaw in equipment or the design of an experiment. The error is reproducible if you experiment in precisely the same manner. In principle, systematic errors can be discovered and corrected, although this may not be easy.

For example, a pH meter that has been standardized incorrectly produces a systematic error. Suppose you think the pH of the buffer used to standardize the meter is 7.00, but it is 7.08. Then, all your pH readings will be 0.08 pH units too low. When you read a pH of 5.60, the actual pH of the sample is 5.68. This systematic error could be discovered by using a second buffer of known pH to test the meter.

Systematic error is a consistent error that can be detected and corrected. Another systematic error arises from an uncalibrated burette. The manufacturer's tolerance for a Class A 50-mL burette is ±0.02 mL. When you think you have delivered 29.43 mL, the real volume could be anywhere from 29.41 to 29.45 mL and still be within tolerance. One way to correct an error of this type is to construct a calibration curve. To do this, deliver distilled water from the burette into a flask and weigh it. Determine the volume of water from its mass by density.

A key feature of systematic error is that it is reproducible. Systematic error may be positive in some regions and consistently negative in others. With care and cleverness, you can detect and correct a systematic error.

2.2.2 Random Error

Random error, indeterminate error, arises from the measurement's uncontrolled (and maybe uncontrollable) variables. A random error has an equal chance of being positive or negative. It is always present and can not be corrected. There is a random error associated with reading a scale. People read the scale to report values representing their subjective interpolation between the markings. One person reading the same instrument several times might report several different readings. Another random error results from electrical noise in a device. Positive and negative fluctuations occur at an approximately equal frequency and cannot be eliminated.

2.2.3 Precision and Accuracy

Precision describes the reproducibility of a result. Your measurement is precise if you measure a quantity several times and the values agree closely. If the values vary widely, your size is not exact. *Accuracy describes how close a measured value is to the "true" value. If a known standard is available, accuracy is how relative your value is to the known value.*

Some standards laboratories in other countries sell certified reference materials, such as clinical and environmental standards and engineering materials, that you can use to test the accuracy of your analytical procedures. The quantity of analyte in a reference material is certified—with meticulous care—to lie in a stated range.

A measurement might be reproducible but wrong. For example, suppose you made a mistake preparing a solution for titration. In that case, you might do a series of reproducible titrations but report an incorrect result because the concentration of the titrating solution was not what you intended. In this case, precision is good, but accuracy is poor. Conversely, it is possible to make poorly reproducible measurements clustered around the correct value. In this case, precision is poor, but accuracy is good. *An ideal procedure is both precise and accurate.*

Accuracy is defined as nearness to the "true" value. True is in quotes because somebody must measure the "true" value, and every measurement has an error. *The "true" value is best obtained by an experienced person using a well-tested procedure.* Testing the result using different techniques is desirable because the systematic error could lead to poor agreement between methods. *Among several ways, a good deal affords us confidence but never proves that results are accurate.*

2.2.4 Absolute and Relative Uncertainty

Absolute uncertainty expresses the margin of uncertainty associated with a measurement. For example, if the estimated uncertainty in reading a calibrated burette is ± 0.02 mL, we say that ± 0.02 mL is the absolute uncertainty associated with the reading.

Relative uncertainty compares the size of the absolute uncertainty with its associated measurement. The relative uncertainty of a burette reading of (12.35 ± 0.02) mL is a dimensionless quotient:

$$\text{Absolute uncertainty} = x - \mu \qquad \textbf{(Equation 2.2)}$$

$$\text{Relative uncertainty} = \frac{\text{Absolute uncertainty}}{\text{Magnitude of measurement}} = \frac{0.02\ \text{mL}}{12.35\ \text{mL}} = 0.002 \qquad \textbf{(Equation 2.3)}$$

The percent relative uncertainty is simple.

$$\begin{aligned}\text{Percent relative uncertainty} &= 100 \times \text{relative uncertainty} \\ &= 100 \times 0.002 = 0.2(\%)\end{aligned} \qquad \textbf{(Equation 2.4)}$$

If the absolute uncertainty in reading a burette is constant at ± 0.02 mL, the percent relative uncertainty is 0.2% for a volume of 10 mL and 0.1% for a volume of 20 mL.

2.3 Statistics

All measurements contain experimental error, so it is impossible to be completely certain of a result.

Example 2.1

Is my red blood cell count high today?

Nevertheless, we often seek the answers to questions such as "Is my red blood cell count today higher than usual?" Of course, if today's count is twice as high as usual, it is probably higher than normal. But what if the "high" count is not excessively above "normal"?

Count on normal days—today's count.

$$\left.\begin{matrix}5.1\\5.3\\4.8\\5.4\\5.2\end{matrix}\right\}\times 10^6\ \text{cells}/\mu\text{L} \qquad 5.6\times 10^6\ \text{cells}/\mu\text{L}$$

Normal values=5.16.

The number 5.6 is higher than the five normal values, but the random variation in normal values might lead us to expect that 5.6 will be observed on some "normal" days..

We will learn in example 2.1 that there is only a 8.5% random chance of observing a value as far from the average as 5.6 on a "normal" day. It is still up to you to decide what to do with this information.

Experimental measurements always contain some variability, so no conclusion can be drawn with certainty. Statistics gives us tools to accept findings with a high probability of being correct and reject conclusions that do not.

2.3.1 Gaussian Distribution

If an experiment is repeated many times and the errors are random, then the results tend to cluster symmetrically about the average value (Figure 2.2). *The more times the experiment is repeated, the more closely the results approach an ideal smooth curve called the Gaussian distribution.* In general, we cannot make so many measurements in a lab experiment. *We are more likely to repeat an experiment 3-5 times than 2 000.* However, we can estimate parameters that describe the large group from the small set of results. We can then make estimates of statistical behavior from the small number of measurements.

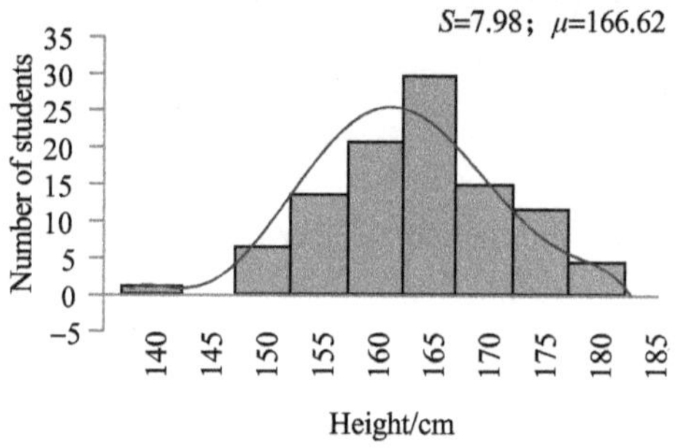

Figure 2. 2 Bar graph and Gaussian curve describing the average height of students. The smooth curve shows the same mean, standard deviation, and area as the bar graph. However, any finite data set will differ from the bell-shaped curve. The more measurements we make, the closer the results will come to the smooth curve

2.3.2 Mean Value and Standard Deviation

In the hypothetical case, a manufacturer tested the average height of 120 students. The bar graph shows the average size of students in each 5-cm interval. The average height approximates a Gaussian distribution because variations in students' construction, such as genetic material and living conditions, are random. The smooth curve is the Gaussian distribution that best fits the data. Any finite set of data will vary somewhat from the Gaussian curve.

The average height and the corresponding Gaussian curve are characterized by two values divided by n, the number of measurements:

Mean: $$\overline{x} = \frac{\sum_i x_i}{n}$$ **(Equation 2. 5)**

where x_i is the lifetime of an individual bulb. The Greek capital sigma, $\sum$, means summation: $\sum_i x_i = x_1 + x_2 + x_3 + \cdots + x_n$. In Figure 2. 2, the mean value is 166. 6 cm.

The deviation, d_i, measures how closely the data between x_i and $\overline{x}$.

Deviation: $$d_i = x_i - \overline{x}$$ **(Equation 2. 6)**

The relative mean deviation, $\overline{d}$, measures how closely.

Mean deviation: $$\overline{d} = \frac{\sum_{i=1}^{n} |d_i|}{n} = \frac{\sum_{i=1}^{n} |x_i - \overline{x}|}{n}$$ **(Equation 2. 7)**

The relative mean deviation, $\overline{d_r}$, measures how closely the data.

Relative mean deviation: $\bar{d}_r = \frac{\bar{d}}{\bar{x}} = \frac{\sum_{i=1}^{n} |x_i - \bar{x}|}{n \times \bar{x}}$ **(Equation 2. 8)**

The standard deviations measure how closely the data are clustered around the mean. The smaller the standard deviation, the more closely the data are clustered around the mean.

Standard deviation: $S = \sqrt{\frac{\sum_i (x_i - \bar{x})^2}{n-1}}$ **(Equation 2. 9)**

In Figure 2. 2, S=7. 98 cm. A set of students having a small standard deviation in life is more uniformly manufactured than a set with a large standard deviation.

The mean for an infinite data set is designated by the lowercase Greek letter mu, μ (the population mean), and the standard deviation is written as a lowercase Greek sigma, σ (the population standard deviation). We can never measure σ, but the values of $\bar{x}$ and s approach σ as the number of measurements increases.

Population standard deviation: $\sigma = 2 - \sqrt{\frac{\sum_i (x_i - \mu)^2}{n}}$ **(Equation 2. 10)**

The quantity n−1 in Equation 2. 9 is called ***the degrees of freedom***. The square of the standard deviation is called the variance. The standard deviation is expressed as a percentage of the mean value ($=100 \times S/\bar{x}$) is called *the relative standard deviation* or the *coefficient of variation*.

Example 2. 2

Find the average, standard deviation, and coefficient of variation for 821, 783, 834, and 855.

The average is

$$\bar{x} = \frac{821+783+834+855}{4} = 823.2$$

Solution:

To avoid accumulating round-off errors, retain one more digit than was present in the original data. The standard deviation is

$$S = \sqrt{\frac{(821-823.2)^2+(783-823.2)^2+(834-823.2)^2+(855-823.2)^2}{4-1}} = 30.3$$

The average and the standard deviation should end at the same decimal place. For $\bar{x}$ = 823. 2, we will write S=30. 3. The coefficient of variation is the percent relative uncertainty:

$$\text{Coefficient of variation} = 100 \times \frac{S}{\bar{x}} = 100 \times \frac{30.3}{823.2} = 3.7\%$$

2.3.3 Significant Figures in Mean and Standard Deviation

We commonly express experimental results in the form $\overline{x} \pm S$ (n=?), where n is the number of data points. It is sensible to write the preceding result as (823 ± 30) (n=4) or even $(8.2 \pm 0.3) \times 10^2$ (n=4) to indicate that the mean has just two significant figures. The expressions (823 ± 30) and $(8.2 \pm 0.3) \times 10^2$ are not suitable for continued calculations in which "$\overline{x}$" and "S" are intermediate results. Retain one or more insignificant digits to avoid introducing round-off errors into subsequent work.

The standard deviation measures the width of the Gaussian curve. The larger the value of σ, the broader the curve. In any Gaussian curve, 68.3% of the area is in the range from $(\mu - 1\sigma)$ to $(\mu + 1\sigma)$. More than two-thirds of the measurements are expected to lie within one standard deviation of the mean. Also, 95.5% of the area lies within $(\mu \pm 2\sigma)$, and 99.7% of the area lies within $(\mu \pm 3\sigma)$. Suppose that you use two different techniques to measure sulfur in coal: method A has a standard deviation of 0.4%, and method B has a standard deviation of 1.1%. You can expect that approximately two-thirds of measurements from method A will lie within 0.4% of the mean. For method B, two-thirds will lie within 1.1% of the mean.

2.4 Confidence Intervals

Student's t-test is a statistical tool used most frequently to express confidence intervals and to compare results from different experiments. For example, it is the tool you could use to evaluate the probability that your red blood cell count will be found in a certain range on "normal" days.

2.4.1 Calculating Confidence Intervals

From a limited number of measurements (n), *we cannot find the true population means*, μ, *or the true standard deviation*, σ. *What we determine are* "$\overline{x}$" and "S", *the sample mean, and the sample standard deviation.* The confidence interval is computed from the equation:

$$\text{Confidence interval} = \overline{x} \pm \frac{tS}{\sqrt{n}} \qquad \textbf{(Equation 2.11)}$$

where t is Student's t, taken from Table 2.1 for the desired confidence level, such as 95%. The meaning of the confidence interval is this: If we were to repeat the n measurements many times to compute the mean and standard deviation, the 95% confidence interval would include the true population mean (whose value we do not know) in 95% of the sets of n measurements. We say (somewhat imprecisely) that "we are 95% confident that the true mean lies within the confidence interval."

Table 2.1 Values of Student's test

Degrees of freedom (df)	Confidence level/%						
	50	90	95	98	99	99.5	99.9
1	1.00	6.31	12.71	31.82	63.66	127.32	636.58
2	0.82	2.92	4.30	6.97	9.93	14.09	31.60
3	0.77	2.35	3.18	4.54	5.84	7.45	12.92
4	0.74	2.13	2.78	3.75	4.60	5.60	8.61
5	0.73	2.02	2.57	3.37	4.03	4.77	6.87
6	0.72	1.94	2.45	3.14	3.71	4.32	5.96
7	0.71	1.90	2.37	3.00	3.50	4.03	5.41
8	0.71	1.86	2.31	2.90	3.36	3.83	5.04
9	0.70	1.83	2.26	2.82	3.25	3.69	4.78
10	0.70	1.81	2.23	2.76	3.17	3.58	4.59
15	0.69	1.75	2.13	2.60	2.95	3.25	4.07
20	0.69	1.73	2.09	2.53	2.85	3.15	3.85
∞	0.67	1.65	1.96	2.33	2.58	2.81	3.29

Example 2.3

Calculating confidence intervals

The carbohydrate content of a glycoprotein (a protein with sugars attached to it) is found to be 12.6%, 11.9%, 13.0%, 12.7%, and 12.5% (g carbohydrate/100 g glycoprotein) in replicate analyses. Find the 50% and 90% confidence intervals for the carbohydrate content.

Solution:

First, calculate $\overline{x}$ (=12.54) and S (=0.40) for the five measurements. For the 50% confidence interval, look up t in Table 2.1 under 50 and across from four degrees of freedom (degrees of freedom = $n-1$). The value of t is 0.74, so the 50% confidence interval is:

$$50\%\ \text{confidence interval} = \overline{x} \pm \frac{tS}{\sqrt{n}}$$

$$= 12.54 \pm \frac{0.74 \times 0.40}{\sqrt{5}} = (12.54 \pm 0.13)\%$$

The 90% confidence interval is:

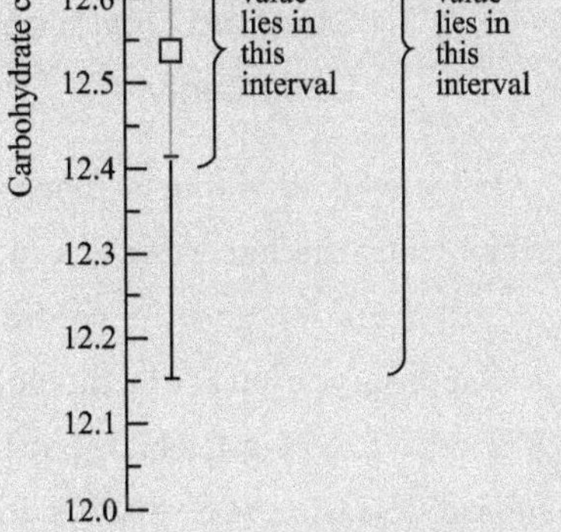

Figure 2.3 The meaning of the 50% and 90% confidence intervals for the carbohydrate content

$$90\%\ \text{confidence interval} = \bar{x} \pm \frac{tS}{\sqrt{n}} = 12.54 \pm \frac{2.13 \times 0.40}{\sqrt{5}} = (12.54 \pm 0.38)\%$$

If we repeated sets of five measurements many times, half of the 50% confidence intervals are expected to include the true mean, μ. Nine-tenths of the 90% confidence intervals are expected to have the true mean, μ.

2.4.2 Meaning of a Confidence Interval

Figure 2.3 illustrates the meaning of confidence intervals. A computer randomly chose numbers from a Gaussian population with a population mean μ of 166.62 cm and a population standard deviation σ of 7.94 in Equation 2.9. In trial 1, three numbers were chosen, and their mean and standard deviation were calculated with Equations 2.5 and 2.9. The 50% confidence interval was then calculated with Equation 2.11, using $t=0.816$ from Table 2.1 (50% confidence, 2 degrees of freedom). This trial is plotted as the first point at the left in Figure 2.4A; the square is centered at the mean value of 166.6, and the error bar extends from the lower limit to the upper limit of the 50% confidence interval ($\pm$ 0.82). The experiment was repeated 105 times to produce the points in Figure 2.4.

The 50% confidence interval is defined such that if we repeated this experiment an infinite number of times, 50% of the error bars in Figure 2.4 would include the true population means of 166.62 cm. we experimented 100 times, and 45 error bars in Figure 2.4 pass through the horizontal line at 166.6 cm.

Figure 2.4 shows the same experiment with the same set of random numbers, but the 90% confidence interval was calculated this time. For an infinite number of experiments, we expect 90% of the confidence intervals to include the population mean of 166.62 cm. Figure 2.4 shows 7 of the 100 error bars crossing the horizontal line at 166.62 cm.

2.4.3 Standard Deviation and Confidence Interval as Estimates of Experimental Uncertainty

If we were dividing a mass by a volume to find density, the uncertainty in density is derived from the uncertainties in mass and volume. Typical estimates of uncertainty are the standard deviation and the confidence interval.

Suppose you measure the volume of a vessel five times and observe values of 6.375 mL, 6.372 mL, 6.374 mL, 6.377 mL, and 6.375 mL. The average is $\bar{x}=6.374\ 6$ mL, and the standard deviation is $S=0.001\ 8$ mL. We could report a volume of (6.374 6 $\pm$ 0.001 8) mL ($n=5$), where n is the number of measurements.

You could choose a confidence interval (such as 95%) to estimate uncertainty. Using Equation 2.9 with 4 degrees of freedom, you find that the 95% confidence interval is $\pm\frac{tS}{\sqrt{n}}=$

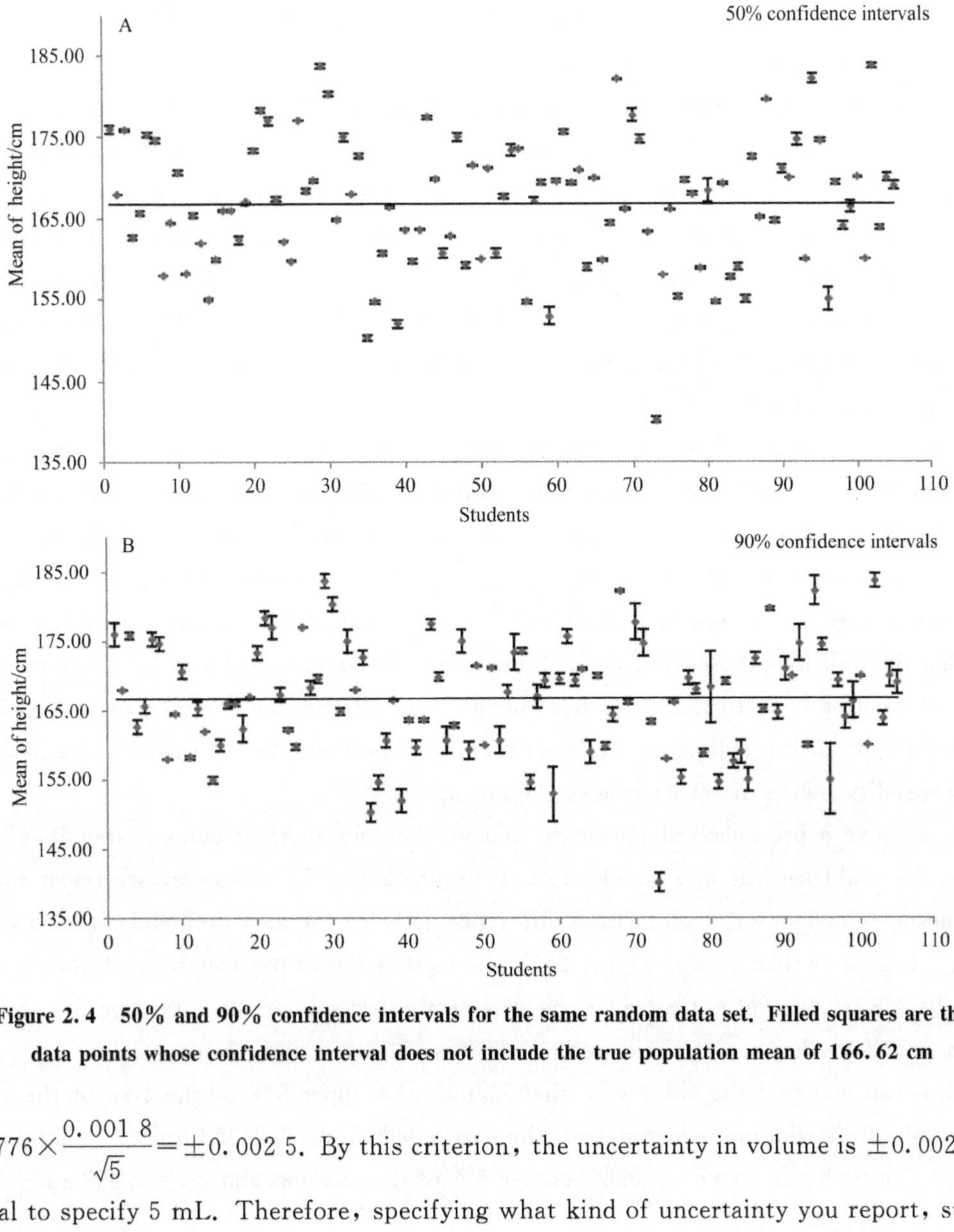

Figure 2.4 50% and 90% confidence intervals for the same random data set. Filled squares are the data points whose confidence interval does not include the true population mean of 166.62 cm

$\pm 2.776 \times \frac{0.0018}{\sqrt{5}} = \pm 0.0025$. By this criterion, the uncertainty in volume is ± 0.002. It is critical to specify 5 mL. Therefore, specifying what kind of uncertainty you report, such as the standard deviation for n measurements or the 95% confidence interval for n measurements, is critical.

We reduce uncertainty by making more measurements. If we make 21 measurements and have the same mean and standard deviation, the 95% confidence interval is reduced from ± 0.0025 to $\pm 2.086 \times \frac{0.0018}{\sqrt{21}} = \pm 0.0008$ mL.

2.5 Comparison of Means with Student's Test

If you make two sets of measurements of the same quantity, the mean value from one set will generally not be equal to that of the other set because of small, random variations in the measurements. We use a t-test to compare one mean value with another to decide whether the two have a statistically significant difference. That is, do the two means agree "within experimental error"? The null hypothesis in statistics states that the mean values from two sets of measurements are not different. Statistics give us a probability that the observed difference between two means arises from random measurement error. We customarily reject the null hypothesis if there is less than a 5% chance that the practical difference arises from random variations. This criterion gives us a 95% chance that our conclusion is correct. One time out of 20, we will be wrong when we conclude that the two means are not different.

One-tailed and two-tailed significance tests. In Equation 2. 11, we sought to compare the mean from four replicate measurements with a certified value. The curve in Figure 2. 5A is the t distribution for 3 degrees of freedom. Suppose the certified value lies under the curve's outer 5% area. In that case, we reject the null hypothesis and conclude with 95% confidence that the measured mean is not equivalent to the certified value. The critical value of t for rejecting the null hypothesis is 3. 18 for 3 degrees of freedom in Table 2. 1. In Figure 2. 5A, 2. 5% of the area beneath the curve lies above $t=3.182$ and 2. 5% below $t=-3.182$. We call this a two-tailed test because we reject the null hypothesis if the certified value lies in the low probability region on either side of the mean.

If we have a preconceived reason to believe that our method gives systematically low values, we could use the one-tailed t-test in Figure 2. 5B. In this case, we reject the null hypothesis (which states no significant difference between the measured and certified values) if $t_{\text{caculated}}$ is greater than 2. 353. Figure 2. 5B shows that 5% of the area beneath the curve lies above $t=2.353$. We did not consider the area on the left side of the curve because we had a reason to believe that our method produces low results, not high results.

How can you find the value of t that bounds the upper 5% of the area of the curve? Because the t distribution is symmetric, the two-tailed value $t=2.353$ for 90% confidence in Table 2. 1 must be the value we seek because 5% of the area lies above $t=2.353$ and 5% lies below $t=-2.353$.

This discussion aims to explain the distinction between one-tailed and two-tailed tests. All t-tests in this book will be two-tailed

At the opening of this chapter, red cell counts on five "normal" days were 5.1×10^6 cells/μL, 5.3×10^6 cells/μL, 4.8×10^6 cells/μL, 5.4×10^6 cells/μL, and 5.2×10^6 cells/μL. The question was whether today's 5.6×10^6 cells/μL count is "significantly" higher than normal. Disregarding the factor of 10^6. the mean of the normal values is $\bar{x}=5.16$ and the standard deviation is $S=0.23$. For today's value of 5. 6.

$$t_{\text{calculated}}=\frac{|x-\bar{x}|}{S}\sqrt{n}=\frac{|5.16-5.6|}{0.23}\sqrt{5}=4.28$$

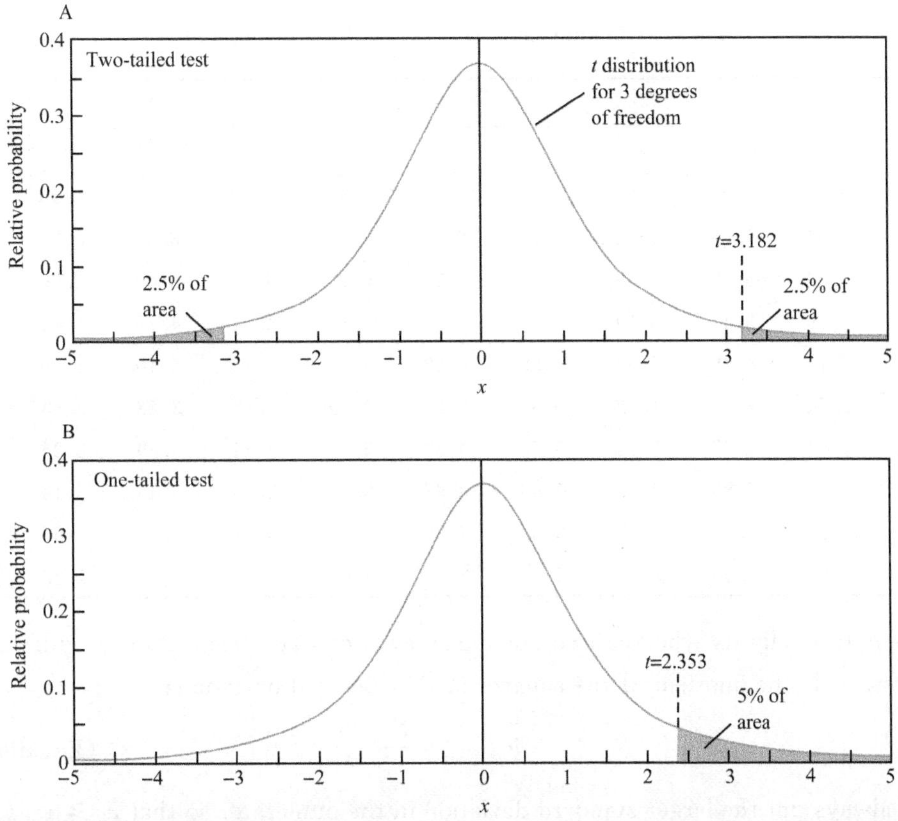

Figure 2. 5 Student's *t* distribution for 3 degrees of freedom. In panel A, each shaded tail contains 2. 5% of the area under the curve. In panel B, the single shaded tail contains 5% of the area. The fewer the degrees of freedom, the broader the distribution. As the degrees of freedom increase, the curve approaches a Gaussian curve

What is the probability of finding $t=4.28$ for 4 degrees of freedom?

In Table 2. 1, looking across the row for 4 degrees of freedom, we see that 4. 28 lies between the 98% ($t=3.75$) and 99% ($t=4.60$) confidence levels. Today's red cell count lies in the upper tail of the curve, containing less than 2% of the area of the curve. There is less than a 2% probability of observing a count of 5.6×10^6 cells/μL on "normal" days. It is reasonable to conclude that today's count is elevated.

2. 6 Comparison of Standard Deviations with the *F* Test

To decide whether Rayleigh's two sets of nitrogen masses in Figure 2. 6 are "significantly" different from each other, we used the t-test. If the standard deviations of the two data sets are not significantly different, then we use Equation 2. 11 for the t-test. However, if the standard deviations are very different, we use Equation 2. 12 instead.

Critical values of F for a one-tailed test of the hypothesis that $S_1>S_2$, there is a 5% probability of observing F above the tabulated value.

Table 2.2 Critical values of $F=\frac{S_1^2}{S_2^2}$ at 95% confidence level

f_2	f_1									
	2	3	4	5	6	7	8	9	10	∞
2	19.00	19.20	19.20	19.30	19.30	19.40	19.40	19.40	19.40	19.50
3	9.55	9.28	9.12	9.01	8.94	8.89	8.84	8.81	8.79	8.53
4	6.94	6.59	6.39	6.26	6.16	6.09	6.04	6.00	5.96	5.63
5	5.79	5.41	5.19	5.05	4.95	4.88	4.82	4.77	4.74	4.36
6	5.14	4.76	4.53	4.39	4.28	4.21	4.15	4.10	4.06	3.67
7	4.74	4.35	4.12	3.97	3.87	3.79	3.73	3.68	3.64	3.23
8	4.46	4.07	3.84	3.69	3.58	3.50	3.44	3.39	3.35	3.93
9	4.26	3.86	3.63	3.48	3.37	3.29	3.23	3.18	3.14	2.71
10	4.10	3.71	3.48	3.33	3.22	3.14	3.07	3.02	2.98	2.54
∞	3.00	2.60	2.27	2.21	2.10	2.01	1.94	1.88	1.83	1.00

The F test tells us whether two standard deviations are "significantly" different from each other. F is the quotient of the squares of the standard deviations:

$$F_{\text{calculated}}=\frac{S_1^2}{S_2^2} \qquad \textbf{(Equation 2.12)}$$

We always put the larger standard deviation in the numerator so that $F \geqslant 1$. Then, we test the hypothesis that $S_1 > S_2$ using the one-tailed F test in Table 2.2. If $F_{\text{calculated}} > F_{\text{table}}$, then the difference is significant.

Example 2.4

Is the standard deviation from chemical decomposition significantly greater than the standard deviation from air in Rayleigh's data?

In Table 2.2, the standard deviation from chemical decomposition is $S_1=0.001\ 38$ ($n_1=8$ measurements), and the standard deviation from air is $S_2=0.000\ 143$ ($n_2=7$ measurements).

Solution:

To answer the question, find F with Equation 2.12:

$$F_{\text{calculated}}=\frac{S_1^2}{S_2^2}=\frac{0.001\ 38^2}{0.001\ 43^2}=93.1$$

In Table 2.2, look for F_{table} in the column with 7 degrees of freedom for S_1 (because of degrees of freedom $=n-1$) and the row with 6 degrees for s_2. Because $F_{\text{caculated}}$ ($=93.1$) $>$ F_{table} ($=4.21$), we accept the hypothesis that $S_1 > S_2$ above the 95% confidence level. The obvious difference in the scatter of the two data sets in Figures 2.9 is highly significant.

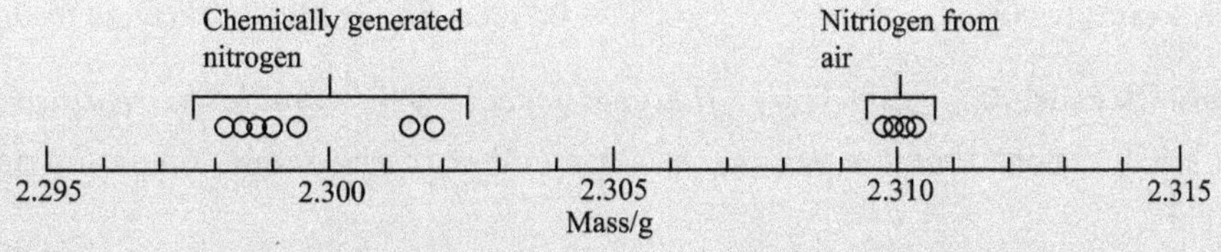

Figure 2.6 Lord Rayleigh's measurements of the mass of constant volumes of gas (at constant temperature and pressure) isolated by removing oxygen from air or generated by decomposition of nitrogen compounds. Rayleigh recognized that the difference between the two clusters was outside of his experimental error and deduced that a heavier component, which turned out to be argon, was present in gas isolated from air

2.7 Test for an Outlier

Students dissolved zinc from a galvanized nail and measured the mass lost to tell how much of the nail was zinc. Here are 12 results:

Mass loss (%): 10.2, 10.8, 11.6, 9.9, 9.4, 7.8, 10.0, 9.2, 11.3, 9.5, 10.6, 11.6

The value 7.8 appears out of line. *A datum that is far from other points is called an outlier.* Should 7.8 be discarded before averaging the rest of the data, or should 7.8 be retained?

2.7.1 Q-Test for an Outlier

We answer this question with the Q test. First, compute the difference between adjacent data ($|x_{\text{adjacent}}| = 1.4$) and the peak-to-peak value ($x_{p\text{-}p} = 3.8$) of the complete data set (all 12 points in this example). Then compute the Q statistic Q, defined as:

Q test: $$Q_{\text{calculated}} = \frac{|X_{\text{questionable value}} - x_{\text{adjacent}}|}{X_{\max} - x_{\min}}$$ **(Equation 2.13)**

The numerator is the absolute difference value between the suspected outlier and the mean value. The questionable point should be discarded if the $Q_{\text{calculated}}$ is greater than Q in Table 2.3.

Table 2.3 Critical values of Q for rejection of outlier

n	0.9	0.95	0.99	n	0.9	0.95	0.99
3	0.90	0.97	0.99	9	0.44	0.51	0.60
4	0.76	0.84	0.93	10	0.41	0.49	0.57
5	0.64	0.73	0.82	11	0.39	0.47	0.54
6	0.56	0.64	0.74	12	0.37	0.46	0.52
7	0.51	0.59	0.68	13	0.36	0.45	0.50
8	0.47	0.54	0.63				

In our example, $Q_{calculated}=\frac{|7.8-11.6|}{|11.6-7.8|}=0.368$. In Table 2.3, $Q_{table}=0.46$ for 12 observations. Because $Q_{calculated}<Q_{table}$, *the questionable point should be retained*. There is more than a 5% chance that the value 7.8 is a member of the same population as the other measurements.

2.7.2 Grubbs Test for an Outlier

We also answer this question using the Grubbs test. First, compute the average ($\overline{x}=10.16$) and the standard deviation ($S=1.11$) of the complete data set (all 12 points in this example). Then compute the Grubbs statistic G, defined as:

Grubbs test: $$G_{calculated}=\frac{\text{Questionable value}-\overline{x}}{S}$$ **(Equation 2.14)**

The numerator is the absolute difference value between the suspected outlier and the mean value. The questionable point should be discarded if $G_{calculated}$ is greater than G in Table 2.4.

In our example, $G_{calculated}=\frac{|7.8-10.16|}{1.11}=2.13$. In Table 2.4, $G_{table}=2.285$ for 12 observations. Because $G_{calculated}<G_{table}$, *the questionable point should be retained*. There is more than a 5% chance that the value 7.8 is a member of the same population as the other measurements.

Common sense must prevail. If you know that a result is low because you spilled part of a solution, then the probability of a false result is 100%, and the datum should be discarded. Any data based on a faulty procedure should be discarded, no matter how well it fits the rest of the data.

Table 2.4 Critical values of *G* for rejection of outlier

n	0.9	0.95	0.975	0.99	0.995
3	1.148	1.153	1.155	1.155	1.155
4	1.425	1.463	1.481	1.492	1.496
5	1.602	1.672	1.715	1.749	1.764
6	1.729	1.822	1.877	1.944	1.975
7	1.828	1.938	2.020	2.097	2.139
8	1.909	2.032	2.126	2.221	2.274
9	1.977	2.110	2.215	2.323	2.387
10	2.036	2.176	2.290	2.410	2.482
11	2.088	2.234	2.355	2.485	2.564
12	2.134	2.285	2.412	2.550	2.636
13	2.175	2.331	2.462	2.607	2.699
14	2.213	2.371	2.507	2.659	2.755
15	2.247	2.409	2.549	2.705	2.806

Key Glossary

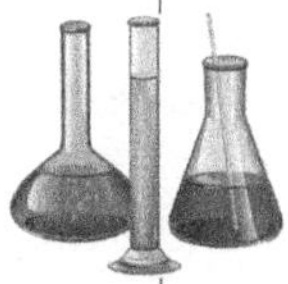

Confidence level is a statistical measure used to express the degree of certainty that a result or finding is accurate. It is often presented as a percentage and is used in hypothesis testing and estimation.

Significant figures also known as significant digits, are the digits in a number that carry meaning in terms of precision. They are important in scientific and engineering calculations as they indicate a measurement's accuracy level.

Error is a deviation from the expected or desired outcome. Errors can be caused by various factors, including syntax, logical, or runtime errors.

Confidence intervals are a statistical tool used to estimate the range in which a population parameter, such as a mean or proportion, is likely to fall. They provide a measure of the uncertainty inherent in statistical estimates, considering both the sample size and the variability of the data.

Student's *t*-test is a statistical method used to determine if there is a significant difference between the means of two groups. It is commonly used in research and experimental studies to compare the averages of two data sets and determine if any observed differences are statistically significant.

Standard deviation is a statistical measure representing the variation or dispersion in a set of values. It indicates how much individual data points differ from the mean of the set. A low standard deviation suggests that the data points are close to the mean, while a high standard deviation indicates that the data points are spread out over a wider range.

Systematic error refers to a consistent deviation from the true value in a measurement process. This type of error is caused by flaws in the measurement system or the environment in which the measurement is taken.

Random error refers to the unpredictable fluctuations in data that occur due to chance and cannot be controlled. It can be caused by various factors such as environmental conditions, measurement errors, or human error.

Precision refers to the level of exactness and consistency in the results obtained. It reflects the ability of a measurement or analysis to provide consistent and accurate results without significant variation or error.

Accuracy refers to the closeness of a measured or calculated value to its true value. It is an important concept in various fields, including science, engineering, and statistics, as it directly impacts the reliability and validity of results.

Essential Exercises

Review Questions

1. Noun explains

Standard deviations, Systematic error, Random error, Precision, Accuracy, Absolute uncertainty, Relative uncertainty

2. Are the absolute values of the error and the absolute error the same?

3. Why should the average value be reported when reporting the analysis results?

4. Which error will be caused by each of the following situations in the analysis process? How do we reduce the error?

(1) The rusted weight.

(2) When the sample is weighed, it absorbs water from the air.

(3) The last digit of the read burette is inaccurate.

(4) Precipitation penetrates the filter paper.

(5) The reagent has a small amount of the analyte.

(6) The sample is not well mixed.

5. Explain random and systematic errors, precision, and accuracy, then mean the relationship between them.

6. What is a blank test? Which error can be reduced by a blank test?

7. How many digits are there in the following significant digits?

(1) 0.010 5, (2) 0.014%, (3) 3.6×10^{-6},

(4) 1 000, (5) pH=4.7, (6) $pK_a=4.74$

8. Calculate the correct number of significant figures.

(1) $9.15\div4.970$

(2) $\dfrac{2.290\times10^6}{6.7\times10^4}$

(3) $17.6-2.838+2.3+110.77$

(4) $(9\,443+45-9.9)\times8.1\times10^{-3}$

Cumulative Problems

1. If the two standard deviations from chemical decomposition were $S_1^2=(0.002\,00)^2$ (7 degrees of freedom) and $S_2^2=(0.001\,00)^2$ (6 degrees of freedom), is the difference significant?

2. Find the average, standard deviation, and coefficient of variation for 410.5, 391.5, 417.0, and 427.5.

3. Researchers measured the carbohydrate content in a sample in parallel six times, with an average of 12.3% and a standard deviation of 0.12%. Please calculate a 90% confidence interval.

Chapter 3

Introduction to Titrimetric Analysis

Key Points

- *Titrimetric Analysis*
- *Titration Analysis Methods*
- *Standard Solution*
- *Calculation of Titration Analysis*
- *Errors in Titration Analysis*

Titrations are crucial in analytical chemistry for determining various substances such as acids, bases, oxidants, reductants, metal ions, and proteins. This method relies on the reaction between the substance being analyzed (analyte) and a standardized reagent known as the titrant, with a known and reproducible stoichiometry. The volume or mass of the titrant required to react with the analyte completely is measured and used to calculate the quantity of the analyte. A figure demonstrates a volume-based titration, where the standard solution is added from a burette, and the reaction takes place in an Erlenmeyer flask. The chemical equivalence point, known as the end point, is indicated by a color change in the indicator or a change in an instrumental response. This chapter provides an overview of titration principles, including the calculations for determining the quantity of the unknown. Additionally, it introduces titration curves and their application in subsequent chapters. Subsequent chapters delve into specific types of titrations, such as neutralization, precipitation, complexometric titrations, and redox titrations.

3.1 Overview

Titration analysis, also known as **volumetric analysis**, involves adding a known and precise concentration of reagent solution, also known as **standard solution** or **titrant**, drop by drop into the solution to be tested in a volumetric flask using a burette until the added reagent reacts completely with the substance being tested according to the stoichiometric relationship. Then, the method calculates the content of the tested component based on the concentration of the reagent solution and the volume added. This process of adding the standard solution drop by drop to the tested solution is called **titration.**

When the standard solution is titrated with the substance being tested in a chemical stoichiometric relationship, it is considered to have reached the **stoichiometric point**, also known as the **titration point**. Since many reactions do not exhibit obvious visual changes at the stoichiometric point, adding an **indicator** to the test solution is often necessary to determine the stoichiometric point based on the indicator's color change. The point at which the indicator changes color is known as the **endpoint of the titration**. However, the endpoint of the titration may not necessarily coincide with the stoichiometric point but rather be in the vicinity of it. The difference between the two is known as the **endpoint error**, which is one of the main sources of error in titration analysis and can generally be controlled within a range of ±0.1% to ±0.2%. The magnitude of the endpoint error mainly depends on the extent to which the titration reaction is complete and whether the choice of indicator is appropriate.

3.1.1 Characteristics of Titration Analysis Methods

Titration analysis methods are widely used in analytical chemistry to determine the concentration of a specific substance in a sample. These methods involve the gradual addition of a reagent of known concentration to the sample until a chemical reaction is completed. The key characteristics of titration analysis methods include their high precision, accuracy, and ability to quantify a wide range of substances, from acids and bases to metals and pharmaceuticals. Additionally, titration methods are relatively simple and easily automated, making them popular for routine laboratory analysis. Titration analysis methods offer a versatile and reliable approach to quantifying substances in various samples. Its characteristics are summarized as follows:

(1) Applicability The titration analysis method is mainly used to determine the content of components above 1%.

(2) Accuracy It has high accuracy, with a relative error of less than 0.1%, making the measurement results of the titration analysis method very reliable. However, it has low sensitivity.

(3) Convenience The operation process of the titration analysis method is simple, easy to master and apply.

(4) Economy The instruments and equipment used are simple and cost-effective, suitable for analysis laboratories of different scales.

(5) Reproducibility The method has good repeatability and reproducibility, ensuring measurement stability.

(6) Diversity The titration analysis method has many applications and can be used to determine various types of chemical reactions.

3.1.2 Requirements for Titration Reactions in Titration Analysis

Not all chemical reactions are suitable for titration analysis due to specific criteria that must be met.

(1) The reactions used for titration analysis must strictly adhere to a specific stoichiometric

relationship to ensure accurate quantitative determination. Any deviation from this relationship can compromise the precision of the titration analysis.

(2) The reaction should exhibit high completeness, with significant changes in solution properties near the stoichiometric point. This ensures the indicator is highly sensitive to color changes and minimizes endpoint errors. To control endpoint error within a plus or minus 0.1% range, the reaction's completeness between the tested substance and the standard should be 99.9% or higher.

(3) The reaction must be rapid and instantaneous. In cases where the reaction is slow, methods such as heating or adding catalysts can be employed to accelerate the process.

(4) There must be suitable methods to determine the endpoint, such as adding an indicator or utilizing techniques like potentiometric titration or photometric titration. These stringent conditions ensure the reliability and accuracy of titration analysis in chemical processes.

3.1.3 Methods of Titration Analysis

Following established standards, titration analysis can be classified into four distinct methods based on the nature of the solution and the specific negative reaction being measured.

The first method, **acid-base titration**, referred to as neutralization titration, relies upon proton transfer reactions. This method has widespread applications in agriculture for assessing the acidity and alkalinity of various agricultural samples and analyzing the presence of nitrogen, phosphorus, carbonates, sulfates, and other substances.

The second method, **precipitation titration**, employs precipitation titration reactions to ascertain the presence of halide ions and silver salts in samples.

The third method, **complexometric titration**, is rooted in complexation reactions and is utilized for determining metal ions such as calcium, magnesium, zinc, copper, and nickel, as well as sulfates and phosphates.

Lastly, **redox titration**, the fourth method, makes use of redox reactions for titration analysis and can be applied to ascertain the presence of potassium, calcium, iron, and organic matter in soil fertilizers, arsenic, and copper in pesticides, as well as reducing sugars and vitamin C in biochemical analysis. These methods collectively form the foundation of titration analysis applications.

3.1.4 Approach to Conducting Titration Analysis

Titration analysis encompasses several methods for determining the concentration of a substance in a solution.

(1) **Direct titration** is the most commonly used method involving using a standard solution to titrate the analyte directly. This method applies to chemical reactions that satisfy certain conditions, such as the reaction between HCl and NaOH or $K_2Cr_2O_7$ and Fe^{2+}.

However, there are instances where direct titration is not feasible, prompting the use of alternative methods.

(2) **Back titration** is employed when the reaction between the analyte and titrant is slow or when a solid sample cannot be immediately titrated to completion. Additionally, back titration is utilized when suitable indicators are lacking or when the analyte interferes with the indicator. In this approach, an excess of a standard solution is first added to the sample, allowing the reaction to proceed. The remaining excess is then titrated with another standard solution. For example, in the titration of Al^{3+}, an excess of EDTA standard solution is first added, and the remaining EDTA is back-titrated using standard Zn^{2+}, Pb^{2+}, or Cu^{2+} solutions. When measuring the content of limestone, whose main component is calcium carbonate ($CaCO_3$), a known amount of excess hydrochloric acid (HCl) standard solution can be added to the sample. After the HCl fully reacts with $CaCO_3$, the remaining HCl can be dripped back into the standard sodium hydroxide (NaOH) solution to calculate the $CaCO_3$ content in the limestone.

$$CaCO_3(\%)=\frac{[c(\mathrm{HCl})V(\mathrm{HCl})-c(\mathrm{NaOH})V(\mathrm{NaOH})]\frac{1}{2}M(CaCO_3)}{m_{\mathrm{sample}}}$$

(3) Another method is **displacement titration**, which involves allowing a substance to react with another compound to displace a certain amount of the substance that can be titrated. For instance, sodium thiosulfate (NaS_2O_3) cannot directly titrate potassium dichromate ($K_2Cr_2O_7$) or other strong oxidizing agents due to lacking a definite stoichiometric relationship. However, NaS_2O_3 can be used as a titrant for iodine (I_2). In this case, an excess of potassium iodide (KI) is added to a solution of $K_2Cr_2O_7$, and the $K_2Cr_2O_7$ displaces a certain amount of I_2, which is then directly titrated with a standard NaS_2O_3 solution.

$$Cr_2O_7^{2-}+6I^-+14H^+ = 2Cr^{3+}+3I_2+7H_2O$$
$$I_2+2S_2O_3^{2-} = 2I^-+S_4O_6^{2-}$$

(4) Finally, **indirect titration** involves determining the concentration of a substance through a separate chemical reaction. For instance, the determination of calcium (Ca) using potassium permanganate ($KMnO_4$) is an example of indirect titration. Since calcium ions (Ca^{2+}) do not undergo redox reactions, they cannot be directly titrated using an oxidation-reduction method. However, by precipitating calcium as calcium oxalate (CaC_2O_4), dissolving it in sulfuric acid (H_2SO_4), and titrating the oxalate ions with $KMnO_4$, the calcium content can be indirectly determined.

$$Ca^{2+}+C_2O_4^{2-} = CaC_2O_4\downarrow$$
$$CaC_2O_4+H_2SO_4 = H_2C_2O_4+CaSO_4$$
$$2MnO_4^-+5H_2C_2O_4+6H^+ = 2Mn^{2+}+10CO_2\uparrow+8H_2O$$

In conclusion, the application of back titration, displacement titration, and indirect titration has significantly expanded the scope of titration analysis, allowing for a wider range of applications.

3.2 Standard Solution

A **standard solution** is a known concentration solution commonly used in analytical chemistry experiments as a reference or calibration material. The concentration and composition of the standard solution are precisely determined, which can help determine the concentration or purity of unknown substances. In titration analysis, it is essential to use standard solutions to calculate the content of the component being tested based on the concentration and volume of the standard solution. Therefore, the correct preparation and proper storage of standard solutions are significant in improving the accuracy of the analytical results.

3.2.1 Methods of Representing Standard Solution Concentration

Two common methods are usually used to express the concentration of a standard solution.

1. Molarity

Molarity, aka molar concentration, is a term that describes the proportion of moles of solute in a solution to the volume of the solution. It is commonly denoted in moles per liter. This measurement is crucial in understanding the chemical composition of a solution and its properties.

$$c_B(\text{mol} \cdot \text{L}^{-1}) = \frac{n_B(\text{mol})}{V_B(\text{L})}$$

2. Titer

Titer, the concentration of a substance in each sample as determined by titration, is a term utilized in the laboratory setting to ascertain the amount of a particular substance in a solution. It is commonly employed in serum or viral solutions to specify the concentration or composition of a specific substance. Titre is expressed in grams per milliliter ($g \cdot mL^{-1}$) of standard solution containing a standard substance, with the symbol $\boldsymbol{T}_s$ (s represents the chemical formula of the standard substance), or represents the mass of each milliliter of standard solution equivalent to the tested substance in grams, with the symbol $\boldsymbol{T}_{s/x}$ (x represents the chemical formula of the tested substance).

Assuming A is the standard solution and B is the test solution, and if the following chemical reaction occurs in the test solution B determined by standard solution A, the conversion relationship between titration and concentration is as follows:

$$a\text{A} + b\text{B} = c\text{C} + d\text{D}$$

$$T_{s/x}=T_s\frac{aM_B}{bM_A}$$

$$T_{s/x}=c_s\times10^{-3}\times\frac{b}{a}\times M_B$$

$$T_s=c_s\times10^{-3}\times M_A$$

For instance, if 1 L of sodium hydroxide (NaOH) solution contains 40.00 g of NaOH, the titration of NaOH is $T(\text{NaOH})=0.040\ 00\ \text{g}\cdot\text{mL}^{-1}$. Suppose this solution is used to determine the content of hydrochloric acid (HCl). In that case, it can also be represented by the mass of the tested substance HCl, i.e., the titer $T(\text{NaOH/HCl})$ of NaOH solution to HCl.

$$T(\text{NaOH/HCl})=T(\text{HCl})\frac{M(\text{HCl})}{M(\text{NaOH})}=0.040\ 00\times\frac{36.46}{40.00}=0.036\ 46(\text{g}\cdot\text{mL}^{-1})$$

3.2.2 Methods of Preparing a Standard Solution

A standard solution is a reagent solution with a precisely known concentration, often used as a titrant in titration analysis. There are two methods for preparing standard solutions: the **direct method** and the **indirect method.**

1. Direct method

The **direct method of preparing a standard solution** involves a series of precise steps to accurately determine a given substance's concentration. Initially, a specific quantity of the reference substance is meticulously weighed to ensure precision. Subsequently, the substance is dissolved in a suitable solvent, and the resulting solution is quantitatively transferred to a volumetric flask. Once the solution has been transferred, it is diluted to the mark on the flask, with thorough shaking to ensure homogeneity. Finally, the accurate concentration of the standard solution is calculated. Figure 3.1 illustrates this method is crucial in analytical chemistry and other fields where precise measurements are paramount. The meticulous nature of each step in this procedure underscores the importance of strict protocols in preparing standard solutions, ensuring the reliability and accuracy of subsequent analyses and experiments.

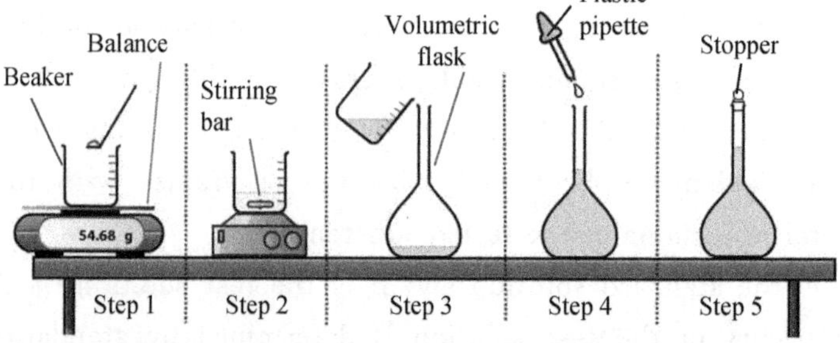

Figure 3.1 Steps for preparing standard solutions using direct methods

For example, if we want to make 250.0 mL of a 0.050 0 mol · L^{-1} solution of sodium carbonate:

(1) Step 1: calculate and weigh the correct quantity of the substance.

The first step of making a standard solution is to calculate the quantity of the substance we want to make the solution out of.

$$n = c \times V$$

$$n(Na_2CO_3) = 0.050\,0 \times 0.250 = 0.012\,5(\text{mol})$$

$$m = n \times M$$

$$m(Na_2CO_3) = 0.012\,5 \times 106.0 = 1.33(\text{g})$$

Therefore, 1.33 g of sodium carbonate is required to make this solution.

(2) Step 2: dissolve the solid.

From step 1, 1.33 g of sodium carbonate is dissolved in a small amount of distilled water in a beaker of appropriate size. Use a stirring rod to help completely dissolve the solid.

(3) Step 3: transfer the standard solution to a volumetric flask.

Transfer the solution to a 250.0-mL volumetric flask using a funnel. Ensure that the volumetric flask is rinsed with distilled water before use. It is appropriate to clean the volumetric flask with water because distilled water will eventually be added to the flask to make the standard solution. Ensure complete transfer of sodium carbonate by rinsing the beaker, stirring the rod, funneling it with distilled water, and discarding the rinsing into the volumetric flask.

(4) Step 4: add distilled water until the required volume.

Once the solution is transferred to the volumetric flask, add distilled water until the level reaches 1 cm below the graduated mark of the flask.

Use a plastic pipette to add distilled water dropwise until the bottom of the meniscus is levelled with the graduated mark.

(5) Step 5: homogenize the solution.

Stopper the flask and invert it up to 10 times.

The substance used to prepare standard solutions directly is known as **a primary standard or primary reagent.** A primary standard is a highly purified compound that serves as reference material in titrations and other analytical methods. The accuracy of a method critically depends on the properties of the primary standard. Important requirements for a primary standard are the following:

(1) High purity. The reagent typically requires a purity of 99.9% or higher, with such low impurity content that it can be disregarded.

(2) Accurately known chemical composition. The material composition must fully conform to the chemical formula. Absence of hydrate water so that the solid composition does not change with variations in humidity.

(3) High stability, resistant to changes during configuration and storage, such as drying

without decomposition, not easily absorbing water during weighing, not reacting with oxygen and carbon dioxide in the air, and not easily deteriorating.

(4) Higher molar weight (variations in mass have a reduced effect on moles). The greater the molar mass, the more mass is weighed, leading to a decrease in weighing error.

Few chemical reagents fully meet these conditions. Even for primary standards that have met the conditions, some treatment is usually necessary before use. The most common treatment is drying the water at a specific temperature. Table 3.1 presents commonly used primary standards, drying adjustments, and application ranges.

Table 3.1 Drying adjustment and application scope of commonly used primary standard

Name	Chemical formula	Composition after drying	Drying conditions and temperature	Name of a calibration object
Sodium bicarbonate	$NaHCO_3$	Na_2CO_3	270-300 ℃	Acid
Anhydrous sodium carbonate	Na_2CO_3	Na_2CO_3	180-200 ℃	Acid
Sodium carbonate decahydrate	$Na_2CO_3 \cdot 10H_2O$	$Na_2CO_3 \cdot 10H_2O$	270-300 ℃	Acid
Potassium bicarbonate	$KHCO_3$	K_2CO_3	270-300 ℃	Acid
Sodium oxalate	$Na_2C_2O_4$	$Na_2C_2O_4$	130 ℃	Oxidant
Oxalic acid dihydrate	$H_2C_2O_4 \cdot 2H_2O$	$H_2C_2O_4 \cdot 2H_2O$	Room temperature air drying	Base or $KMnO_4$
Borax	$Na_2B_4O_7 \cdot 10H_2O$	$Na_2B_4O_7 \cdot 10H_2O$	Placed in a dryer containing saturated solutions of NaCl and sucrose	Acid
Potassium hydrogen phthalate	$KHC_8H_4O_4$	$KHC_8H_4O_4$	110-120 ℃	Acid
Potassium dichromate	$K_2Cr_2O_7$	$K_2Cr_2O_7$	140-150 ℃	Reductant
Potassium bromate	$KBrO_3$	$KBrO_3$	130 ℃	Reductant
Potassium iodate	KIO_3	KIO_3	130 ℃	Reductant
Copper	Cu	Cu	Store in a room-temperature dryer	Reductant
Arsenic trioxide	As_2O_3	As_2O_3	Store in a room-temperature dryer	Oxidant
Calcium carbonate	$CaCO_3$	$CaCO_3$	105-110 ℃	EDTA
Zinc	Zn	Zn	Store in a room-temperature dryer	EDTA
Zinc oxide	ZnO	ZnO	900-1 000 ℃	EDTA
Sodium chloride	NaCl	NaCl	500-600 ℃	Silver nitrate
Benzoic acid	$C_7H_6O_2$	$C_7H_6O_2$	Sulfuric acid vacuum dryer	Sodium methoxide, etc
p-Aminobenzenesulfonic acid	$C_6H_7O_3NS$	$C_6H_7O_3NS$	120 ℃	Sodium nitrite

Many substances used to prepare standard solutions do not meet the above conditions. For example, sodium hydroxide (NaOH) readily absorbs carbon dioxide (CO_2) and water from the air. Hydrochloric acid (HCl) is volatile, making it challenging to determine its precise HCl content. Potassium permanganate ($KMnO_4$), sodium thiosulfate ($Na_2S_2O_3$), and other substances are not easily purified and can decompose when exposed to light. As a result, standard solutions for these substances can only be prepared indirectly.

2. Indirect method

The **indirect preparation** of a standard solution involves the rough weighing of a specific substance or solution, followed by the preparation of a solution with a concentration roughly close to the required level. The accurate concentration is then determined using a standard solution of a primary standard or another substance, known as calibration. This step is crucial in ensuring the accuracy and reliability of the standard solution for use in various analytical procedures.

There are two methods for calibrating standard solutions.

(1) **Direct calibration method.** An accurate amount of standard substance is measured, dissolved, and then titrated with the standard solution to be calibrated. Then, based on the mass of the standard substance and the volume of the standard solution consumed, the accurate concentration of the standard solution is calculated. Most standard solutions are determined for their accurate concentration using this method. For example, to prepare a $0.1\ mol \cdot L^{-1}$ HCl standard solution, first dilute concentrated HCl to make a solution of approximately $0.1\ mol \cdot L^{-1}$, then accurately measure a certain amount of standard substance such as borax, dissolve it, and titrate it with the diluted HCl solution to be calibrated until both react completely. Then, calculate its accurate concentration based on the volume of HCl solution consumed in the titration.

(2) **Comparative titration method.** It's a way of accurately determining the concentration of a solution by comparing it to a known concentration standard solution. You compare the volumes of two solutions and their known concentrations to calculate the accurate concentration of the solution you want to determine. This method isn't as accurate as direct titration, so in order to distinguish it, the standard solution used in comparative titration is called a secondary standard. Only primary standards, which are directly titrated, are used for analytical work that requires high accuracy. The standard solution must be stored in tightly sealed reagent bottles (or plastic bottles) to prevent evaporation and dust contamination. Some standard solutions also need protection from CO_2 in the air, light-induced oxidation, or decomposition, as these factors can change the concentration of the solution. Long-term stored standard solutions should be periodically re-titrated based on specific circumstances. Before each use, the standard solution should be thoroughly mixed to prevent concentration changes caused by partial evaporation and condensation on the upper part of the bottle. Standard solutions easily decomposed by light, such as $AgNO_3$ and

$KMnO_4$, should be stored in brown bottles and kept in a dark place. Strong alkaline solutions that can absorb CO_2 from the air and corrode glass are best stored in plastic bottles, with a soda lime (Na_2CO_3+CaO) tube attached to the bottle mouth to absorb CO_2 and water from the air.

3.3 Calculation of Titration Analysis

The calculation of titration analysis includes the preparation and calibration of the solution, the quantitative relationship between the titrant and the analyte, and the calculation of the analysis results.

3.3.1 Calculation Related to Solution Preparation

(1) When diluting from a concentrated solution to a diluted solution, the amount of substance before and after dilution remains unchanged, and the following equation can be used to calculate:

$$c_{\text{before dilution}} V_{\text{before dilution}} = c_{\text{after dilution}} V_{\text{after dilution}}$$

(2) When preparing a solution from a solid substance, the following equations can be used to calculate:

$$c=\frac{n}{V};\quad n=\frac{m}{M};\quad c=\frac{m}{VM};\quad m=c\times V\times M$$

Example 3.1

To prepare a 500.0 mL 0.1000 mol · L^{-1} $AgNO_3$ solution by direct method, how many grams of pure $AgNO_3$ should be taken? How should it be prepared?

Solution:

The mass of pure $AgNO_3$ should be assumed to be m.

$$m=c\times V\times M$$

$$m=0.1000\times 500\times 10^{-3}\times 169.9=8.495(\text{g})$$

Using an analytical balance, accurately weigh 8.495 g of pure $AgNO_3$ and place it in a clean 100 mL beaker. Add a suitable amount of pure water to dissolve it, then transfer the solution quantitatively to a 500-mL volumetric flask and shake it thoroughly after making up the volume. Store the solution in a dry brown reagent bottle away from light. Alternatively, rinse a non-dry brown reagent bottle with a small amount of $AgNO_3$ solution three times, then pour the $AgNO_3$ solution into the bottle to prevent dilution and changes in concentration.

Example 3.2

To prepare 1 L of 0.10 mol · L^{-1} HCl solution. How many milliliters of concentrated HCl should be taken from concentrated HCl (relative density 1.18 containing 37% HCl)?

Solution:

(1) Calculate the molar concentration of concentrated HCl.

Based on the calculation relationship:

$$c_B=\frac{n_B}{V}=\frac{m_B}{M_B V}=\frac{m_B}{M_B\frac{m}{\rho}}=\frac{\rho\frac{m_B}{m}}{M_B}=\frac{w_B\rho}{M_B}$$

Substitute the known data:

$$c_{\text{before dilution}}=\frac{1.18\times 37\%\times 1\,000}{36.5}\approx 12(\text{mol}\cdot\text{L}^{-1})$$

(2) Calculate the required volume of concentrated HCl.

Based on the calculation relationship:

$$c_{\text{before dilution}}V_{\text{before dilution}}=c_{\text{after dilution}}V_{\text{after dilution}}$$

Substitute the known data:

$$12\times V_{\text{before dilution}}=1\times 0.10$$

$$V_{\text{before dilution}}=8.3(\text{mL})$$

3.3.2 Calculation Relationship Between the Quantities of Reactants

In order to find the quantitative relationship between the quantities of reactants, the reaction coefficient ratio can be determined based on the stoichiometric relationship in the reaction equation to determine the relationship between the quantities of substances. For any titration reaction:

$$a\text{A}+b\text{B}=c\text{C}+d\text{D}$$

$$\frac{n_A}{n_B}=\frac{a}{b};\quad n_A=\frac{a}{b}n_B;\quad n_B=\frac{b}{a}n_A$$

For example, the reaction between $K_2Cr_2O_7$ and Fe^{2+} in acidic solution is:

$$6Fe^{2+}+Cr_2O_7^{2-}+14H^+ = 6Fe^{3+}+2Cr^{3+}+7H_2O$$

$$\frac{n(Fe^{2+})}{n(K_2Cr_2O_7^{2-})}=\frac{6}{1}$$

For example, in acidic solutions, the reaction between $KMnO_4$ and $K_2C_2O_4$:

$$2MnO_4^-+5C_2O_4^{2-}+16H^+ = 2Mn^{2+}+10CO_2+8H_2O$$

$$\frac{n(MnO_4^-)}{n(C_2O_4^{2-})}=\frac{2}{5}$$

In addition, you can choose the basic unit according to habit and then calculate it according to the rules of equivalent quantities of substances. In the four types of reactions in analytical chemistry, the reaction coefficients for the precipitation titration of silver and the EDTA coordination titration are both 1 : 1, which is relatively simple. In acid-base titration, the basic unit of the reactants can be determined based on the transfer of 1 mol or the equivalent transfer of 1 mol of protons. For examples:

$$2NaOH+H_2SO_4 = Na_2SO_4+2H_2O$$

In the reaction, the basic unit of NaOH remains NaOH, while the basic unit of H_2SO_4 is 1/2 H_2SO_4, so $n(NaOH)=n(1/2H_2SO_4)$.

$$Na_2CO_3+2HCl = 2NaCl+H_2O+CO_2$$

In the reaction, the basic unit of HCl remains HCl, while the basic unit of Na_2CO_3 is $1/2Na_2CO_3$, so $n(HCl)=n(1/2Na_2CO_3)$.

In redox titration, the basic unit of the reactant can be determined based on the gain or loss of 1 mol of electrons or the equivalent of gaining or losing 1 mol of electrons. For example:

$$MnO_4^-+5Fe^{2+}+8H^+ = Mn^{2+}+5Fe^{3+}+4H_2O$$

In the reaction, 1 mol of MnO_4^- gains 5 mol of electrons, so the basic unit of MnO_4^- is $1/5MnO_4^-$. The basic unit of Fe^{2+} remains Fe^{2+}. So $n(1/5MnO_4^-)=n(Fe^{2+})$.

Example 3.3

When titrating hydrochloric acid (HCl) with a standard solution of borax ($Na_2B_4O_7 \cdot 10H_2O$) with a concentration of 0.10 mol · L^{-1} How much pure borax should be taken if the titration volume is controlled to around 30 mL?

Solution:

$$2HCl+Na_2B_4O_7 \cdot 10H_2O = 4H_3BO_3+2NaCl+5H_2O$$

$$\frac{n(HCl)}{n(Na_2B_4O_7 \cdot 10H_2O)}=\frac{2}{1}$$

$$c(HCl)\times V(HCl)\times 10^{-3}=2\times\frac{m(Na_2B_4O_7 \cdot 10H_2O)}{M(Na_2B_4O_7 \cdot 10H_2O)}$$

$$m(Na_2B_4O_7 \cdot 10H_2O)=\frac{c(HCl)\times V(HCl)\times 10^{-3}\times M(Na_2B_4O_7 \cdot 10H_2O)}{2}$$

Substitute the known data:

$$m(Na_2B_4O_7 \cdot 10H_2O)=\frac{0.1\times 30\times 10^{-3}\times 381}{2}\approx 0.6(g)$$

Example 3.4

A 0.500 0 g impure sample of K_2CO_3 was titrated with 27.31 mL of 0.106 4 mol·L^{-1} HCl to methyl orange endpoint. Calculate the percentage of K_2CO_3 or K_2O in the sample.

Solution:

$$2HCl + K_2CO_3 = 2KCl + CO_2 + H_2O$$

$$n(K_2CO_3) = n(K_2O) = \frac{1}{2}n(HCl)$$

$$\omega(K_2CO_3) = \frac{\frac{1}{2}c(HCl) \times V(HCl) \times 10^{-3} \times M(K_2CO_3)}{m_{sample}} \times 100\%$$

$$= \frac{0.106\ 4 \times 27.31 \times 10^{-3} \times 138.2}{2 \times 0.500\ 0} \times 100\% = 40.157\ 9\% \approx 40.16\%$$

$$\omega(K_2O) = \frac{\frac{1}{2}c(HCl) \times V(HCl) \times 10^{-3} \times M(K_2O)}{m_{sample}} \times 100\%$$

$$= \frac{0.106\ 4 \times 27.31 \times 10^{-3} \times 94.20}{2 \times 0.500\ 0} \times 100\% = 27.37\%$$

3.3.3 Calculation of Analysis Results

1. Direct titration method

Example 3.5

A 45.00 mL portion of HCl solution required 22.71 mL of 0.021 35 mol·L^{-1} $Ba(OH)_2$ to reach an endpoint with a bromocresol green indicator. Calculate the molar concentration of the HCl.

Solution:

In the titration, 1 mol of $Ba(OH)_2$ reacts with 2 mol of HCl:

$$Ba(OH)_2 + 2HCl \longrightarrow BaCl_2 + H_2O$$

The number of millimoles of the standard is calculated by substituting them into an equation.

$$n(Ba(OH)_2) = V(Ba(OH)_2) \times c(Ba(OH)_2) = 22.71 \times 10^{-3}\ L \times 0.021\ 35\ mol \cdot L^{-1}$$
$$= 0.484\ 8 \times 10^{-3}\ (mol)$$

To find the number of millimoles of HCl, we multiply this result by the stoichiometric ratio determined from the titration reaction:

$$n(HCl) = c(HCl) \times V(HCl)$$

To obtain the number of moles of HCl per litre, we divide by the volume of the acid. Therefore,

$$c(\mathrm{HCl})=\frac{n(\mathrm{Ba(OH)_2})\times 2}{V(\mathrm{HCl})}=\frac{0.484\ 8\ \mathrm{mmol}\times 2}{45.00\ \mathrm{mL}}=0.021\ 55\ \mathrm{mol\cdot L^{-1}}$$

Example 3.6

A titration of 0.220 2 g of pure $Na_2C_2O_4$ (134.00 g · mol^{-1}) required 27.52 mL of $KMnO_4$. What is the molar concentration of the $KMnO_4$ solution? The chemical reaction is:

$$2\mathrm{MnO_4^+}+5\mathrm{C_2O_4^{2-}}+16\mathrm{H^-}\longrightarrow 2\mathrm{Mn^{2+}}+10\mathrm{CO_2}+8\mathrm{H_2O}$$

Solution:

From this equation, we see that

$$n(2\mathrm{KMnO_4})=n(5\mathrm{Na_2C_2O_4})$$

The amount of primary-standard $Na_2C_2O_4$ is

$$n(\mathrm{Na_2C_2O_4})=\frac{m(\mathrm{Na_2C_2O_4})}{M(\mathrm{Na_2C_2O_4})}=\frac{0.220\ 2\ \mathrm{g}}{134.00\ \mathrm{g\cdot mol^{-1}}}$$

To obtain the number of millimoles of $KMnO_4$, we multiply this result by the stoichiometric ratio:

$$n(\mathrm{KMnO_4})=\frac{2}{5}n(\mathrm{Na_2C_2O_4})$$

The concentration of $KMnO_4$ is then obtained by dividing by the volume consumed. Thus,

$$c(\mathrm{KMnO_4})=\frac{n(\mathrm{KMnO_4})}{V(\mathrm{KMnO_4})}=\frac{2}{5}\times\frac{n(\mathrm{Na_2C_2O_4})}{V(\mathrm{KMnO_4})}=\frac{2}{5}\times\frac{0.220\ 2\ \mathrm{g}}{134.00\ \mathrm{g\cdot mol^{-1}}\times 27.52\ \mathrm{mL}}$$
$$=0.023\ 88\ (\mathrm{mol\cdot L^{-1}})$$

Example 3.7

A 0.725 0 g sample of iron ore is dissolved in acid. The iron is then reduced to Fe^{2+} and titrated with 46.56 mL of 0.019 23 mol · L^{-1} $KMnO_4$ solution. Calculate the results of this analysis in terms of (a) % Fe (55.847 g · mol^{-1}) and (b) % Fe_3O_4 (231.54 g · mol^{-1}).

Solution:

The equation describes the reaction of the analyte with the reagent.

$$\mathrm{MnO_4^-}+5\mathrm{Fe^{2+}}+8\mathrm{H^+}\longrightarrow \mathrm{Mn^{2+}}+5\mathrm{Fe^{3+}}+4\mathrm{H_2O}$$

(1) $n(\mathrm{MnO_4^-})=n(5\mathrm{Fe^{2+}})$

The amount of Fe^{2+} is then given by

$$n(Fe^{2+}) = 5 \times n(MnO_4^-) = 5 \times c(MnO_4^-) \times V(MnO_4^-)$$
$$= 5 \times 0.019\,23\ mol \cdot L^{-1} \times 46.56\ mL = 4.477(mmol)$$

The percent Fe^{2+} is

$$\omega(Fe^{2+}) = \frac{m(Fe^{2+})}{m_{\text{sample of aniron}}} \times 100\% = \frac{n(Fe^{2+}) \times M(Fe^{2+})}{m_{\text{sample of aniron}}} \times 100\%$$
$$= \frac{4.477\ mmol \times 55.847\ g \cdot mol^{-1}}{0.725\,0\ g} \times 100\% = 34.49\%$$

(2) To determine the correct stoichiometric ratio, we note that

$$n(Fe_3O_4) = n(3Fe^{2+}) = \frac{1}{3} n(Fe^{2+})$$

$$\omega(Fe_3O_4) = \frac{m(Fe_3O_4)}{m_{\text{sample of aniron}}} \times 100\% = \frac{n(Fe_3O_4) \times M(Fe_3O_4)}{m_{\text{sample of aniron}}} \times 100\%$$
$$= \frac{1}{3} \times \frac{n(Fe^{2+}) \times M(Fe_3O_4)}{m_{\text{sample of aniron}}} \times 100\%$$
$$= \frac{1}{3} \times \frac{4.477\ mmol \times 231.54\ g \cdot mol^{-1}}{0.725\,0\ g} \times 100\% = 47.66\%$$

2. Back titration

Example 3.8

The phosphorus in a 4.258 g sample of plant food was converted to PO_4^{3+} and precipitated as Ag_3PO_4 by adding 50.00 mL of 0.082 0 $mol \cdot L^{-1}$ $AgNO_3$. The excess $AgNO_3$ was back-titrated with 4.06 mL of 0.062 5 $mol \cdot L^{-1}$ KSCN. Express the results of this analysis in terms of % P_2O_5.

Solution:

The chemical reactions are

$$P_2O_5 + 3H_2O \longrightarrow 2PO_4^{3-} + 6H^+$$
$$2PO_4^{3-} + 6Ag^+ \longrightarrow 2Ag_3PO_4$$
$$Ag^+ + SCN^- \longrightarrow AgSCN$$

The stoichiometric ratios are

$$n(P_2O_5) = n(6Ag^+)$$
$$n(Ag^+) = n(SCN^-)$$
$$n(\text{total amount } AgNO_3) = 6n(P_2O_5) + n(SCN^-)$$

$$n(P_2O_5)=\frac{1}{6}\times[n(\text{total amount } AgNO_3)-n(SCN^-)]$$

$$n(P_2O_5)=\frac{1}{6}\times[50.00\ mL\times0.0820\ mol\cdot L^{-1}-4.06\ mL\times0.0625\ mol\cdot L^{-1}]$$

$$\omega(P_2O_5)=\frac{n(P_2O_5)\times M(P_2O_5)}{m_{sample}}\times100\%=\frac{0.641\times10^{-3} mol\cdot L^{-1}\times141.9\ g\cdot mol^{-1}}{4.258\ g}\times100\%$$

$$=2.14\%$$

Example 3.9

The limestone sample (mainly composed of $CaCO_3$) weighing 0.225 0 g was reacted with 0.105 0 mol · L^{-1} HCl in 40.00 mL, and the remaining NaOH was titrated with 0.100 0 mol · L^{-1}, totaling 15.20 mL. Calculate the percentage of $CaCO_3$ in the sample.

Solution:

$$\omega(CaCO_3)=\frac{[c(HCl)V(HCl)-c(NaOH)V(NaOH)]\frac{1}{2}M(CaCO_3)}{m_{sample}}$$

$$\omega(CaCO_3)=\frac{(0.1050\times0.04000-0.1000\times0.01520)\times\frac{1}{2}\times100.09}{0.2250}\times100\%=59.61\%$$

3. Displacement titration

Example 3.10

The CO in a 20.3 L sample of gas was converted to CO_2 by passing the sample over iodine pentoxide(I_2O_5) heated to 150 ℃:

$$O_2(s)+2CO(g)\longrightarrow 2CO_2(g)$$

The iodine was distilled at this temperature and was collected in an absorber containing 8.25 mL of 0.011 01 mol · L^{-1} $Na_2S_2O_3$.

$$I_2(g)+2S_2O_3^{2-}(aq)\longrightarrow 2I^-(aq)+S_4O_6^{2-}(aq)$$

The excess $Na_2S_2O_3$ was back-titrated with 2.16 mL of 0.009 47 mol · L^{-1} I_2 solution. Calculate the concentration of CO (28.01 g · mol^{-1}) in mg per liter of sample.

Solution:

Based on the two reactions, the stoichiometric ratios are

$$n(I_2O_5)=n(5CO)=n(I_2) \text{ and } n(2S_2O_3^{2-})=n(I_2)$$

We divide the first ratio by the second to get a third useful ratio.

$$n(5CO)=n(2S_2O_3^{2-})$$

This relationship reveals that 5 mmol of CO is responsible for consuming 2 mmol of $Na_2S_2O_3$. The total amount of $Na_2S_2O_3$ is

$$n(\text{the totle of } S_2O_3^{2-}) = c(S_2O_3^{2-}) \times V(S_2O_3^{2-}) = 8.25\ \text{mL} \times 0.011\ 01\ \text{mol} \cdot \text{L}^{-1}$$

The amount of $Na_2S_2O_3$ consumed in the back titration is

$$n(S_2O_3^{2-} \text{ in the back-titration}) = 2.16\ \text{mL}\ I_2 \times 0.009\ 47\ \text{mol} \cdot \text{L}^{-1} I_2 \times 2$$

The number of millimoles of CO can be calculated by using the third stoichiometric ratio:

$$n(CO) = \frac{5}{2} \times [n(\text{the totle of } S_2O_3^{2-}) - n(S_2O_3^{2-} \text{ in the back-titration})]$$

4. Indirect titration method

Example 3.11

Weigh 0.160 0 g of limestone sample, dissolve it in acid, add oxalic acid, precipitate with Ca^{2+} at appropriate acidity to obtain calcium oxalate, filter and wash the precipitate, and dissolve it in sulfuric acid. Titrate 21.08 mL of 0.020 00 mol · L^{-1} potassium permanganate standard solution to the endpoint and determine the calcium content in limestone.

Solution:

$$Ca^{2+} + C_2O_4^{2-} = CaC_2O_4 \downarrow$$

$$CaC_2O_4 + H_2SO_4 = H_2C_2O_4 + CaSO_4$$

$$2MnO_4^- + 5H_2C_2O_4 + 6H^+ = 2Mn^{2+} + 10CO_2 \uparrow + 8H_2O$$

$$\frac{n(Ca^{2+})}{n(MnO_4^-)} = \frac{2}{5}$$

$$m(CaCO_3) = \frac{5 \times c(MnO_4^-) \times V(MnO_4^-) \times 10^{-3} M(CaCO_3)}{2}$$

$$m(CaCO_3) = \frac{5 \times 0.020\ 00 \times 0.021\ 08 \times 100.09}{2} = 0.105\ 5(\text{g})$$

$$\omega(CaCO_3) = \frac{m(CaCO_3)}{m_{\text{sample}}} = \frac{0.105\ 5\ \text{g}}{0.160\ 0\ \text{g}} \times 100\% = 65.93\%$$

3.4 Errors in Titration Analysis

Generally, the margin of error for titration analysis typically falls within the 0.1% to 0.2% range. In order to attain such a high degree of precision, it is crucial to have a comprehensive understanding of the potential error sources in the analysis process and employ effective error reduction methods.

3.4.1 Method Error

The main sources of method error are:

(1) The titration endpoint does not coincide with the equivalence point of the chemical reaction.

(2) The indicator participates in the titration reaction, thus consuming a certain amount of standard solution, for example, if the acid-base indicator is a weak acid or weak base or if the redox indicator itself is a weak oxidizing agent or weak reducing agent.

(3) The standard solution is always in excess when reaching the titration endpoint.

To reduce method errors, choosing a suitable indicator as much as possible before titration is important so that the endpoint and the equivalence point are as consistent as possible. The amount of indicator should not be excessive unless specifically required. Generally, 1-2 drops are sufficient. When approaching the equivalence point, adding the standard solution drop by drop, or even half a drop at a time, is necessary to avoid adding it too quickly.

3.4.2 Weighing Error

When using uncalibrated weights, in most cases, the error of the larger weight is greater than that of the smaller weight. However, if the same weights are consistently used, most of the errors can be offset. If the two arms of the balance are not equal, for example, the right arm is 1/100 shorter than the left arm, then the mass added to the right pan of the balance is 1/100 more than the mass on the left pan, resulting in a 1/100 error in the measurement (or the measured mass = actual mass × 1.01). This error is significant, especially if the balance is used for calibration, as the calibration result will be 1/100 higher. However, when using the same balance to weigh the unknown sample, the error is offset, and the content of the sample is not affected. Therefore, using the same balance and weights for calibration and measurement, most errors can be offset even if the balance arms are not equal and the weights are uncalibrated. Thus, calibration is unnecessary in general analysis if the same balance and weights are used, but calibration is required for more precise analysis. If a balance with a sensitivity of 1/10 000 is used to weigh samples with a mass of 0.2 g or more, the weighing error of the sample can be within 0.1‰.

3.4.3 Error in Measuring Volume

When using uncalibrated burettes, measurement errors in the volume can be introduced. For example, when using the same burette for calibration and sample determination, starting each titration from the "0.00" mark, and using similar volumes of solution, the calibration error of the burette can be offset, so it is advisable to carry out calibration and determination under the same conditions. However, some measurement errors cannot be offset, such as titrating Fe^{2+} with a standard solution of $K_2Cr_2O_7$. $K_2Cr_2O_7$ is the reference substance, and the prepared standard solution does not need to be calibrated, so the calibration error of the burette cannot be completely offset. In back titration operations, the volumes of the solutions used for calibration and determination are often not consistent, and their errors cannot be offset either. In titration analysis, when using burettes, pipettes, and volumetric flasks to measure the volume of a solution, if qualified measuring instruments are used and the operating procedures of the instruments are strictly followed, this error can be controlled within 0.1% to 0.2%.

3.4.4 Operational Negligence or Error

In titration analysis, several common operational errors or mistakes can significantly impact the accuracy of the results. Firstly, inadequate mixing of solutions can lead to uneven distribution of substances. It is crucial to thoroughly shake the prepared standard and sample solutions in volumetric flasks after reaching the desired volume, as failure to do so may result in a portion of the solution being inaccurately represented during titration. Secondly, improper storage of standard solutions can lead to changes in concentration, thereby affecting the accuracy of the titration process. Additionally, inadequate cleaning of instruments can introduce impurities and cause inaccurate volume measurements due to irregular liquid surfaces or water droplets. Therefore, it is essential to ensure that instruments are thoroughly cleaned to prevent concentration changes and to understand which instruments require rinsing with operational liquids. Lastly, operational errors such as air bubbles in the burette, rapid titration speed, incorrect reading methods, or sample spillage outside the receiving container during weighing can all lead to inaccuracies. However, these errors can be avoided by carefully following operational procedures. Understanding and addressing these common mistakes is crucial for ensuring the reliability and precision of titration analysis.

Key Glossary

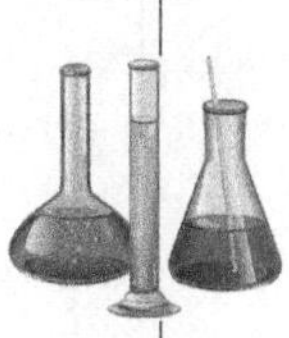

Titration analysis: Titration analysis is a chemical analysis method used to determine the concentration of specific chemical substances in a solution. It is achieved by gradually adding a reagent of known concentration (called the titrant). The titrant reacts with the target compound in the tested solution until the reaction reaches the equivalence point. The concentration of the target compound can be calculated by measuring the titrant used. Titration analysis is commonly used to determine acidity, alkalinity, metal ion content, and the concentration of other chemical substances.

Standard solution: The standard solution is a known concentration solution commonly used in analytical chemistry experiments as a reference or calibration material.

Stoichiometric point: The stoichiometric point in titration refers to the optimal balance point where the added titrant and the substance being titrated reach the best chemical stoichiometric ratio. At this point, the molar ratio of the titrant and the substance being titrated corresponds to the molar ratio in the chemical equation. By reaching the equivalence point, it is possible to determine the moles of the substance being titrated and consequently derive the solute concentration in the solution.

Indicator: A titration indicator is a type of chemical reagent used to help determine the endpoint of a titration. It can indicate changes in a solution's acidity or other properties by changing color or producing specific chemical reactions, thus determining the titration endpoint. Common titration indicators include phenolphthalein, bromothymol blue, and methyl orange.

The endpoint of the titration: The endpoint of a titration refers to the complete state of a chemical reaction, where the indicator changes color or shows other specific signals to indicate the end of the titration.

Endpoint error: The difference between the endpoint of the titration and the stoichiometric point.

Molarity: aka molar concentration, is a term that describes the proportion of moles of solute in a solution to the volume of the solution.

Titer: The concentration of a substance in each sample, as determined by titration, is a term utilized in the laboratory setting to ascertain the amount of a particular substance in a solution.

Primary standard: The primary standard is a highly purified compound that serves as reference material in titrations and other analytical methods, with high purity, accurately known chemical composition, high stability, higher molar weight, and so on.

Essential Exercises

Review Questions

1. What are the concepts of stoichiometric point, titration endpoint, and endpoint error in chemistry?

2. What are the conditions that titration reactions in titration analysis should meet?

3. What is a primary standard? What are the criteria for primary standards? Can the following substances be used as primary standard?

(1) $Na_2CO_3 \cdot 10H_2O$　(2) 99.95% NaCl　(3) HCl　(4) $Na_2B_4O_7 \cdot 10H_2O$

4. Which substances can be prepared into standard solutions by direct method? Which ones can only be prepared by indirect method? If calibration is needed, which reference substances should be chosen?

H_2SO_4,　KOH,　$KBrO_3$,　$K_2Cr_2O_7$,　$Na_2S_2O_3 \cdot 5H_2O$

5. When using anhydrous Na_2CO_3 to standardize HCl solution, what effect does the following situation have on the concentration of the HCl solution?

(1) When calibrating the concentration of HCl solution, a small amount of $NaHCO_3$ is present in the standard substance Na_2CO_3.

(2) Anhydrous Na_2CO_3 absorbs water.

(3) The burette was not rinsed with HCl solution before loading the HCl solution.

(4) When weighing anhydrous Na_2CO_3, the mass of 0.183 4 g was mistakenly recorded as 0.182 4 g.

(5) During weighing, a small amount of distilled water was in the conical flask containing Na_2CO_3.

6. The following substances cannot be used as primary standards ( ).

A. $K_2Cr_2O_7$ B. $Na_2C_2O_4$

C. NaOH D. Potassium hydrogen phthalate

7. Known $T(Na_2C_2O_4/KMnO_4) = 0.006\ 700\ g \cdot mL^{-1}$. What is the concentration of $Na_2C_2O_4$ solution ( )?

A. $0.200\ 0\ mol \cdot L^{-1}$ B. $0.500\ 0\ mol \cdot L^{-1}$

C. $0.106\ mol \cdot L^{-1}$ D. $0.020\ 00\ mol \cdot L^{-1}$

8. The chemical analysis method that calculates the content of the measured component based on the ( ) and ( ) of the standard solution used after titrating the test solution to the endpoint with a standard solution is called titration analysis.

9. The method of titration analysis can be divided into four types based on the different reaction types: ( ), ( ), ( ), and ( ).

10. There are four methods of titration analysis: ( ), ( ), ( ), and ( ).

11. Standard NaOH solution should be prepared indirectly, using ( ) or ( ) as primary standard reagents for titration. When using phenolphthalein and methyl orange as indicators, only ( ) should be used, not ( ).

12. To calibrate the concentration of HCl solution, substance ( ) should be used as the primary standard. Substance ( ) can be used as the primary standard for calibrating the NaOH solution, while substance ( ) can be used for calibrating $KMnO_4$.

13. Using oxalic acid dihydrate $H_2C_2O_4 \cdot 2H_2O$ as the primary substance to standardize the concentration of NaOH solution, but due to improper storage, the oxalic acid dihydrate has lost some of its crystalline water. Using this oxalic acid to standardize the concentration of NaOH solution will yield a ( ) result.

14. When titrating HCl solution with the standard substance Na_2CO_3, the following situations will have what kind of impact on the concentration of HCl (higher, lower, no impact):

(1) If the titration speed is too fast, and the HCl attached to the walls of the burette does not have enough time to flow down before reading the titration volume.

(2) Before pouring the standard HCl solution into the burette, the burette was not rinsed with the HCl solution.

(3) an additional 50 mL of distilled water was added when dissolving in the conical flask.

(4) While weighing Na_2CO_3, a portion was accidentally spilled on the balance pan.

Cumulative Problems

1. Prepare 500 mL of 0.2 mol · L^{-1} H_2SO_4 and 500 mL of 0.2 mol · L^{-1} NaOH. To achieve this, calculate the volume of concentrated H_2SO_4 (density 1.84 g · mL^{-1}, 95% purity) required and determine the mass of solid NaOH needed.

2. What is the titer T(HCl) of the 0.102 4 mol · L^{-1} HCl solution? Also, what is the titer $T(HCl/Na_2CO_3)$ when the HCl solution reacts with Na_2CO_3 to produce CO_2?

3. When titrating with borax ($Na_2B_4O_7 \cdot 10H_2O$) to standardize the concentration of HCl at approximately 0.1 mol · L^{-1}, what is the minimum volume of HCl required to ensure a measurement error of no more than 0.1%? How many grams of borax should be weighed at least? If using Na_2CO_3 as the standardizing agent, how many grams should be considered at least? What is the relative error in the weighing process?

4. The 1.100 g Na_2CO_3 sample requires 35.00 mL of $T(HCl/Na_2CO_3)$=0.026 50 g · mL^{-1} HCl reacts with Na_2CO_3 to produce CO_2 completely. Calculate the sample content.

5. To completely neutralize a mixture of 0.200 g $BaCO_3$ and Na_2CO_3, 30.00 mL of 0.100 mol · L^{-1} HCl is required. Calculate the mass fractions of $BaCO_3$ and Na_2CO_3 in the mixture.

Chapter 4

Acid-Base Titrations

Key Points

- *Visual Indicators*
- *Mixed Indicators*
- *Neutralization Curves*
- *The Determination of a Mixture*

The acid-base titration is a quick and accurate means of determining acidic or basic substances in analytical samples. It determines an acid or base's concentration by neutralizing it with an acid or base of known concentration. Methods based on these are sometimes termed acidimetric or alkalimetric. This allows for quantitative analysis of the concentration of an unknown acid or base solution. Acid-base titration makes use of the neutralization reaction that occurs between acids and bases. A pH titration graph can also determine pK_a and K_a (acid constants). Acid-base titrations can also be used to find the percent purity of chemicals.

4.1 Visual Indicators for Acid-Base Titrations

The object of titrating can be explained as an alkaline solution with a standard solution of an acid, is the determination of the amount of acid, which is equivalent chemically to the amount of base present. The point at which this is reached is the equivalence point, stoichiometric point, or theoretical endpoint. If both the acid and base are strong electrolytes, the solution at the end-point will be neutral and have a pH of 7; but if either the acid or the base is a weak electrolyte, the salt will be hydrolyzed to a certain degree, and the solution at the equivalence point will be either slightly alkaline or slightly acid. The exact pH of the solution at the equivalence point can readily be calculated from the ionization constant of the weak acid or the weak base and the concentration of the solution. For any actual titration, the correct end-point will be characterized by a definite value of the hydrogen-ion concentration of the solution, depending upon the nature of the acid and the base and the concentration of the solution.

Many substances, called **acid-base indicators** or neutralization, change color according to the hydrogen-ion concentration of the solution. The chief characteristic of these indicators is that the change from a predominantly "acid" color to a predominantly "alkaline" color is not sudden and abrupt but takes place within a small interval of pH (usually about two pH units), termed the color-change interval of the indicator. The position of the color-change interval in the pH scale varies widely with different indicators. For most acid-base titrations, selecting an indicator exhibiting a distinct color change at a pH close to the equivalence point is possible.

W. Ostwald suggested the first useful theory of indicator action based on the concept that indicators generally are weak organic acids or bases.

The simple Ostwald theory of the color change of indicators has been revised, and the color changes are believed to be due to structural changes, including the production ofquinonoid and resonance forms; these may be illustrated by reference to phenolphthalein, the changes of which are characteristic of all phthalein indicators: see the Figure 4.1 given below. In the presence of dilute alkali, the lactone ring in 1 opens to yield (b), and the triphenylcarbinol structure (b) undergoes water loss to produce the resonating ion (c), which is red. On the other hand, if phenolphthalein is treated with an excess of concentrated alcoholic alkali, the red color first produced disappears owing to the formation of (d).

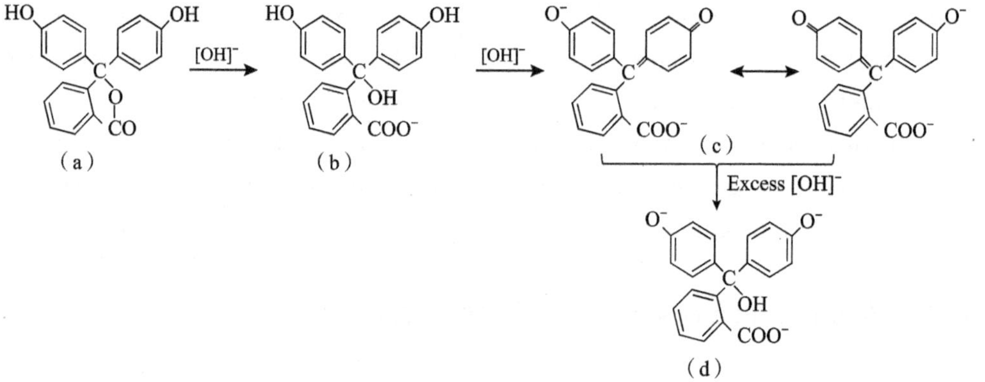

Figure 4.1 The quinonoid and resonance forms changes in phenolphthalein

The Brönsted-Lowry concept of acids and bases makes it unnecessary to distinguish between acid and base indicators: the emphasis is placed on the charge types of the acid and alkaline forms of the indicator. The equilibrium between the acidic form In_A and the basic form In_B may be expressed as:

$$In_A = H^+ + In_B$$

The equilibrium constant is:

$$\frac{\alpha_{H^+} \times \alpha_{In_B}}{\alpha_{In_A}} = K_{In}$$

The ratio of the acidic and basic forms concentrations determines the observed color of an indicator in solution. This is given by:

$$\frac{[In_A]}{[In_B]} = \frac{\alpha_{H^+} \times y_{In_B}}{K_{In} \times y_{In_A}}$$

Where y_{InA} and y_{InB} are the activity coefficients of the acidic and basic forms of the indicator. The equation may be written in the logarithmic form:

$$pH=-\lg a_{H^+}=pK_{In}+\lg\frac{[In_B]}{[In_A]}+\lg\frac{y_{In_B}}{y_{In_A}}$$

The pH will depend upon the ionic strength of the solution (which is, of course, related to the activity coefficient). Hence, when making a color comparison for the determination of the pH of a solution, not only must the indicator concentration be the same in the two solutions, but the ionic strength must also be equal or approximately equal. The equation incidentally explains the so-called salt and solvent effects observed with indicators. A condensed form of the equation can express the color-change equilibrium at any ionic strength (constant activity-coefficient term):

$$pH=pK'_{In}+\lg\frac{[In_B]}{[In_A]}$$

where pK'_{In} is referred to as the apparent indicator constant.

The value of the ratio $[In_B]/[In_A]$ (i. e., [basic form]/[acidic form]) can be determined by a visual color comparison or, more accurately, by a spectrophotometric method. Both forms of the indicator are present at any hydrogen ion concentration. It must be realized, however, that the human eye has a limited ability to detect either of two colors when one of them predominates. Experience shows that the solution will appear to have the "acid" color, i. e., of In_A, when the ratio of $[In_A]$ to $[In_B]$ is above approximately 10, and the "alkaline" color, i. e., of In_B, when the ratio of $[In_B]$ to $[In_A]$ is above about 10. Thus, only the "acid" color will be visible when $[In_A]/[In_B]>10$; the corresponding limit of pH given by the equation is:

$$pH=pK'_{In}-1$$

Only the alkaline color will be visible when $[In_B]/[In_A]>10$, and the corresponding limit of pH is:

$$pH=pK'_{In}+1$$

The color-change interval is accordingly $pH=pK'_{In}\pm 1$, i. e., over approximately two pH units. Within this range, the indicator will appear to change from one color to another. The change will be gradual since it depends upon the ratio of the concentrations of the two colored forms (acidic and basic). When the pH of the solution is equal to the apparent dissociation constant of the indicator pK'_{In}, the ratio $[In_A]$ to $[In_B]$ becomes equal to 1, and the indicator will have color due to an equal mixture of the "acid" and "alkaline" forms. This is sometimes known as the "middle tint" of the indicator. This applies strictly only if the two colors are of equal intensity. If one form is more intensely colored than the other or if the eye is more sensitive to one color than the other, then the middle tint will be slightly displaced along with the pH range of the indicator.

Table 4.1 contains a list of indicators suitable for titrimetric analysis and the colorimetric determination of pH. The color-change intervals of most indicators listed in the table are represented graphically in Figure 4.2.

Table 4.1 Color changes and pH range of certain indicators

Indicator	pH range	Color in acid solution	Color in alkaline solution	pK'_{In}
Brilliant cresyl blue (acid)	0.0-1.0	Red-orange	Blue	—
Cresol red (acid)	0.2-1.8	Red	Yellow	—
m-Cresol purple	0.5-2.5	Red	Yellow	—
Quinaldine red	1.4-3.2	Colorless	Red	—
Thymol blue (acid)	1.2-2.8	Red	Yellow	1.7
Tropacolin OO	1.3-2.8	Red	Yellow	—
Bromophenol blue	2.8-4.6	Yellow	Blue	4.1
Ethyl orange	3.0-4.5	Red	Orange	—
Methyl orange	2.9-4.6	Red	Orange	3.7
Congo red	3.0-5.0	Blue	Red	—
Bromocresol green	3.6-5.2	Yellow	Blue	4.7
Methyl red	4.2-6.3	Red	Yellow	5.0
Ethyl red	4.5-6.5	Red	Orange	—
Chlorophenol red	4.6-7.0	Yellow	Red	6.1
4-Nitrophenol	5.0-7.0	Colorless	Yellow	7.1
Bromocresol purple	5.2-6.8	Yellow	Purple	6.1
Bromophenol red	5.2-7.0	Yellow	Red	—
Azolitmin (litmus)	5.0-8.0	Red	Blue	—
Bromothymol blue	6.0-7.6	Yellow	Blue	7.1
Neutral red	6.8-8.0	Red	Orange	—
Phenol red	6.8-8.4	Yellow	Red	7.8
Cresol red (base)	7.2-8.8	Yellow	Red	8.2
1-Naphtholphthalein	7.3-8.7	Yellow	Blue	8.4
m-Cresol purple	7.6-9.2	Yellow	Purple	—
Thymol blue (base)	8.0-9.6	Yellow	Blue	8.9
O-Cresolphthalein	8.2-9.8	Colorless	Red	—
Phenolphthalein	8.3-10.0	Colorless	Red	9.6
Thymolphthalein	9.3-10.5	Colorless	Blue	9.3
Alizarin yellow R	10.1-12.1	Yellow	Red-orange	—
Brilliant cresyl blue (base)	10.8-12.0	Blue	Yellow	—
Tropacolin OO	11.1-12.7	Yellow	Orange	—
Nitramine	10.8-13.0	Colorless	Orange-brown	—

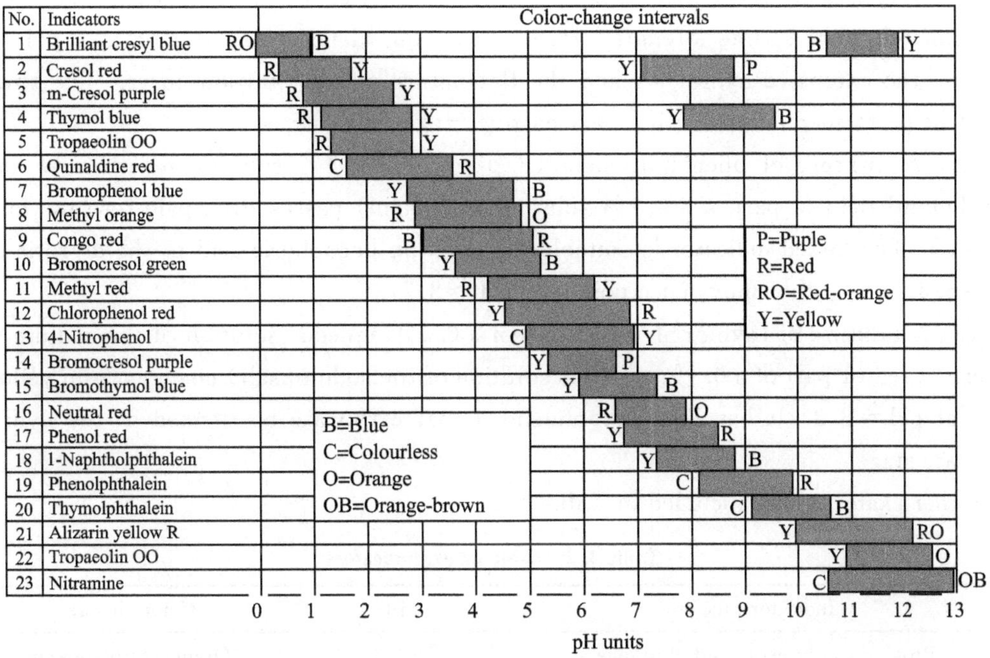

Figure 4. 2 The color-change intervals of most of the various indicators

It is necessary to draw attention to the variable pH of water, which may be encountered in quantitative analysis. For example, water in equilibrium with the normal atmosphere, which contains 0. 03% by volume of carbon dioxide, has a pH of about 5. 7; very carefully prepared conductivity water has a pH close to 7; water saturated with carbon dioxide under a pressure of one atmosphere has a pH of about 3. 7 at 25 ℃. The analyst may, therefore, be dealing, according to the conditions in the laboratory, with water having a pH between the extremes, pH 3. 7 and pH 7. Hence, for indicators that show their alkaline colors at pH values above 4. 5, the effect of carbon dioxide introduced during a titration, either from the atmosphere or from the titrating solutions, must be seriously considered.

4. 2 Mixed Indicators

For some purposes, it is desirable to have a sharp color change over a narrow and selected pH range; this is not easily seen with an ordinary acid-base indicator since the color change extends over two pH units. The required result may, however, be achieved by using a suitable mixture of indicators; these are generally selected so that their pK'_{In} values are close together and the overlapping colors are complementary at an intermediate pH value. A few examples will be given in some detail.

(1) A mixture of equal parts of neutral red (0. 1% solution in ethanol) and methylene blue (0. 1% solution in ethanol) gives a sharp color change from violet-blue to green in passing from acid to alkaline solution at pH 7. This indicator may be employed to titrate

acetic acid (ethanoic acid) with ammonia solution or vice versa. Both acid and base are approximately of the same strength. Hence, the equivalence point will be at a pH ≈ 7, owing to the extensive hydrolysis and the flat nature of the titration curve. The titration cannot be performed except with a very narrow range indicator.

(2) A mixture of phenolphthalein (3 parts of a 0.1% solution in ethanol) and 1-naphtholphthalein (1 part of a 0.1% solution in ethanol) passes from pale rose to violet at pH=8.9. The mixed indicator is suitable for titrating phosphoric acid to the diprotic stage ($K_2=6.3\times10^{-8}$, the equivalence points at pH ≈8.7).

(3) A mixture of thymol blue (3 parts of a 0.1% aqueous solution of the sodium salt) and cresol red (1 part of a 0.1% aqueous solution of the sodium salt) changes from yellow to violet at pH=8.3. It has been recommended that carbonate be titrated to the hydrogen carbonate stage.

Other examples are included in Table 4.2.

Table 4.2 Some mixed indicators

Indicator mixture	pH	Color change
Bromocresol green; methyl orange	4.3	Orange→Blue-green
Bromocresol green; chlorophenol red	6.1	Pale green→Blue violet
Bromothymol blue; neutral red	7.2	Rose pink→Green
Bromothymol blue; phenol red	7.5	Yellow→Violet
Thymol blue; cresol red	8.3	Yellow→Violet
Thymol blue; phenolphthalein	9.0	Yellow→Violet
Thymolphthalein; phenolphthalein	9.9	Colorless→Violet

The color change of a single indicator may also be improved by adding a pH-sensitive dyestuff to produce the complement of one of the indicator colors. A typical example is an addition of xylene cyanol FF to methyl orange (1.0 g of methyl orange and 1.4 g of xylene cyanol FF in 500 mL of 50% ethanol). Here, the color change from the alkaline to the acid side is green + grey + magenta, with the middle (grey) stage at pH=3.8. The above is an example of a screened indicator, and the mixed indicator solution is sometimes known as "screened" methyl orange. Another example is the addition of methyl green (2 parts of a 0.1% solution in ethanol) to phenolphthalein (1 part of a 0.1% solution in ethanol); the former complements the red-violet basic color of the latter, and at a pH of 8.4-8.8 the color change is from grey to pale blue.

4.3 Neutralization Curves

The mechanism of neutralization processes can be understood by studying the changes in the hydrogen ion concentration during the appropriate titration. The difference in pH about the equivalence point is of the most significant importance, as it enables an indicator to be selected, giving the minor titration error. The curve that results from charting pH values on the vertical axis (ordinate) and the percentage of acid that has been neutralized (or the volume of alkali added, in milliliters) on the horizontal axis (abscissa) is referred to as the **neutralization curve**, or more broadly, the **titration curve.** This may be evaluated experimentally by determining the pH at various stages during the titration by a potentiometric method, or it may be calculated from theoretical principles.

4.4 Neutralization of a Strong Acid with a Strong Base

For this calculation, it is assumed that both the acid and the base are completely dissociated. The activity coefficients of the ions are unity to obtain the pH values during the neutralization of the strong acid and the strong base, or vice versa, at the laboratory temperature. For simplicity of calculation, consider the titration of 100 mL of 1 mol • L^{-1} hydrochloric acid with 1 mol • L^{-1} sodium hydroxide solution. The pH of 1 mol • L^{-1} hydrochloric acid is 0. When 50 mL of the 1 mol • L^{-1} the base has been added, 50 mL of unneutralized 1 mol • L^{-1} acid will be present in a total volume of 150 mL.

$[H^+]$ will therefore be $50\times1/150=3.33\times10^{-1}$, or pH=0.5

For 75 mL of base, $[H^+]=25\times1/175=1.43\times10^{-1}$, pH=0.8

For 90 mL of base, $[H^+]=10\times1/190=5.26\times10^{-2}$, pH=1.3

For 98 mL of base, $[H^+]=2\times1/198=1.01\times10^{-2}$, pH=2.0

For 99 mL of base, $[H^+]=1\times1/199=5.02\times10^{-3}$, pH=2.3

For 99.9 mL of base, $[H^+]=0.1\times1/199.9=5.00\times10^{-4}$, pH=3.3

After adding 100 mL of the base, the pH will change sharply to 7, i.e., the theoretical equivalence point. The resulting solution is simply one of sodium chloride. Any sodium hydroxide added beyond this will exceed what is required for neutralization.

With 100.1 mL of base, $[OH^-]=0.1/200.1=5.00\times10^{-4}$, pOH=3.3 and pH=10.7

With 101 mL of the base, $[OH^-]=1/201=5.00\times10^{-3}$, pOH=2.3 and pH=11.7

These results show that as the titration proceeds, initially, the pH rises slowly. Still, between the addition of 99.9 and 100.1 mL of alkali, the pH of the solution rises from 3.3 to 10.7, i.e., in the vicinity of the equivalence point, the rate of change of pH of the solution is very rapid.

The complete results, up to 200 mL of alkali, are collected in Table 4.3; this also includes the figures for 0.1 mol • L^{-1} and 0.01 mol • L^{-1} solutions of acid and base, respectively. The alkali additions have been extended in all three cases to 200 mL; it is evident that the range from 100 to 200 mL and beyond represents the reverse titration of 100 mL of alkali with the acid in the presence of the non-hydrolyzed sodium chloride solution. The data in Table 4. 3 are presented graphically in Figure 4. 3.

Table 4. 3 pH during the titration of 100 mL of HCl with NaOH of equal concentration

V(NaOH) added/ mL	pH		
	1 mol · L^{-1} solution	0. 1 mol · L^{-1} solution	0. 01 mol · L^{-1} solution
0	0. 0	1. 0	2. 0
50	0. 5	1. 5	2. 5
75	0. 8	1. 8	2. 8
90	1. 3	2. 3	3. 3
98	2. 0	3. 0	4. 0
99	2. 3	3. 3	4. 3
99. 5	2. 6	3. 6	4. 6
99. 8	3. 0	4. 0	5. 0
99. 9	3. 3	4. 3	5. 3
100. 0	7. 0	7. 0	7. 0
100. 1	10. 7	9. 7	8. 7
100. 2	11. 0	10. 0	9. 0
100. 5	11. 4	10. 4	9. 4
101	11. 7	10. 7	9. 7
102	12. 0	11. 0	10. 0
110	12. 7	11. 7	10. 7
125	13. 0	12. 0	11. 0
150	13. 3	12. 3	11. 3
200	13. 5	12. 5	11. 5

In quantitative analysis, the changes in pH near the equivalence point are of special interest. This part of Figure 4. 3 is shown on a larger scale in Figure 4. 4, indicating the color-change intervals of some of the common indicators.

With 1 mol · L^{-1} solutions, it is evident that any indicator with an effective range between pH 3 and 10. 5 may be used. As a result, the color change will be sharp, and the titration error will be negligible.

With 0. 1 mol · L^{-1} solutions, the ideal pH range for an indicator is limited to 4. 5-9. 5. Methyl orange will exist chiefly in the alkaline form when 99. 8 mL of alkali have been added, and the titration error will be 0. 2%, which is negligibly small for most practical purposes; it is, therefore, advisable to add sodium hydroxide solution until the indicator is present completely in the alkaline form. The titration error is also negligibly small with phenolphthalein.

With 0. 01 mol · L^{-1} solutions, the ideal pH range is still further limited to 5. 5-8. 5; such indicators as methyl red, bromothymol blue, or phenol red will be suitable. The titration error for methyl orange will be 1-2 percent.

The above considerations apply to solutions that do not contain carbon dioxide. In practice, carbon dioxide is usually present from the small amount of carbonate in the sodium hydroxide and the atmosphere.

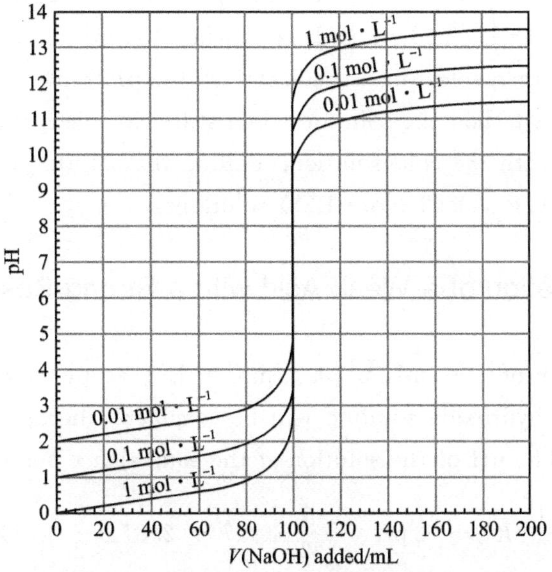

Figure 4. 3　Neutralization curves of 100 mL of HCl with NaOH of the same concentration (calculated)

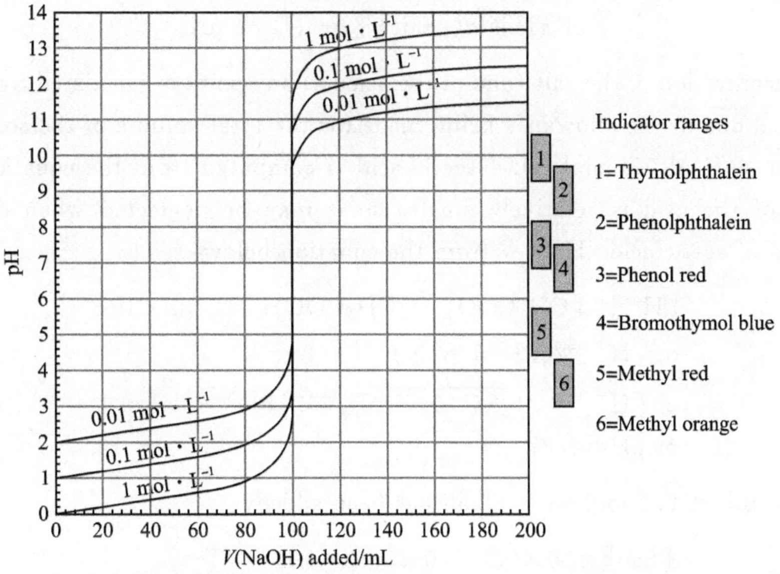

Figure 4. 4　Neutralization curves of 100 mL of HCl with NaOH of the same concentration near the equivalence point (calculated)

The gas is in equilibrium with carbonic acid, which weakens both ionization stages. This will introduce a small error when indicators of a high pH range (above pH 5) are used, e.g., phenolphthalein or thymolphthalein. More acid indicators, such as methyl orange and methyl yellow, are unaffected by carbonic acid. The difference between the amounts of sodium hydroxide solution used with methyl orange and phenolphthalein is not greater than 0.15-0.2 mL of 0.1 mol • L^{-1} sodium hydroxide when 100 mL of 0.1 mol • L^{-1} hydrochloric acid is titrated. Other than selecting an indicator with a pH range below pH 5, a method of eliminating this error is to boil the solution while still acid to expel carbon dioxide and then continue the titration with the cold solution. Boiling the solution is particularly productive when titrating dilute (e.g., 0.01 mol • L^{-1}) solutions.

4.5 Neutralization of a Weak Acid with a Strong Base

The neutralization of 100 mL of 0.1 mol • L^{-1} acetic acid (ethanoic acid) with 0.1 mol • L^{-1} sodium hydroxide solution will be considered here; other concentrations can be treated similarly. The pH of the solution at the equivalence point is given by

$$\mathrm{pH}=\frac{1}{2}\mathrm{p}K_\mathrm{w}+\frac{1}{2}\mathrm{p}K_\mathrm{a}-\frac{1}{2}\mathrm{p}c=7+2.37-\frac{1}{2}(1.3)=8.72$$

For other concentrations, we may employ the approximate mass action expression:

$$[\mathrm{H^+}]\times[\mathrm{CH_3COO^-}]/[\mathrm{CH_3COOH}]=K_\mathrm{a}$$

$$\text{or } [\mathrm{H^+}]=[\mathrm{CH_3COOH}]\times K_\mathrm{a}/[\mathrm{CH_3COO^-}]$$

$$\text{or } \mathrm{pH}=\lg([\mathrm{Salt}]/[\mathrm{Acid}])+\mathrm{p}K_\mathrm{a}$$

The concentration of the salt (and of the acid) at any point is calculated from the volume of alkali added due to the allowance being made for the total volume of the solution.

The initial pH of 0.1 mol • L^{-1} acetic acid is computed from the equation below; the dissociation of the acid is relatively small, so it may be neglected when expressing the concentration of acetic acid. Hence, from the equation below:

$$[\mathrm{H^+}]\times[\mathrm{CH_3COO^-}]/[\mathrm{CH_3COOH}]=1.82\times10^{-5}$$

$$\text{or } [\mathrm{H^+}]^2/0.1=1.82\times10^{-5}$$

$$\text{or } [\mathrm{H^+}]=\sqrt{1.82\times10^{-6}}=1.35\times10^{-3}$$

$$\text{or } \mathrm{pH}=2.87$$

When 50 mL of 0.1 mol • L^{-1} alkali has been added,

$$[\mathrm{Salt}]=50\times0.1/150=3.33\times10^{-2}$$

$$\text{and } [\mathrm{Acid}]=50\times0.1/150=3.33\times10^{-2}$$

$$\mathrm{pH}=\lg(3.33\times10^{-2}/3.33\times10^{-2})+4.74=4.74$$

The pH values at other points on the titration curve are similarly calculated. After the equivalence point has been passed, the solution contains an excess of OH^- ions, which will repress the hydrolysis of the salt; the pH may be assumed, with sufficient accuracy for our purpose, to be that due to the excess of the base present, so that in this region the titration curve will almost coincide with that for 0.1 mol • L^{-1} hydrochloric acid (Figure 4.3 and Table 4.3). All the results are collected in Table 4.4 and are depicted graphically in Figure 4.5. The results for the titration of 100 mL of 0.1 mol • L^{-1} solution of a weaker acid ($K_a = 1\times10^{-7}$) with 0.1 mol • L^{-1} sodium hydroxide at the laboratory temperature is also included.

Table 4.4 Neutralization of 100 mL of 0.1 mol • L^{-1} acetic acid ($K_a=1.82\times10^{-5}$) and 100 mL of 0.1 mol • L^{-1} HA ($K_a=1\times10^{-7}$) with 0.1 mol • L^{-1} sodium hydroxide

Volume of 0.1 mol • L^{-1} NaOH used/mL	0.1 mol • L^{-1} acetic acid ($K_a=1.82\times10^{-5}$) (pH)	0.1 mol • L^{-1} HA ($K_a=1\times10^{-7}$) (pH)
0	2.9	4.0
10	3.8	6.0
25	4.3	6.5
50	4.7	7.0
90	5.7	8.0
99.0	6.7	9.0
99.5	7.0	9.3
99.8	7.4	9.7
99.9	7.7	9.8
100.0	8.7	9.9
100.2	10.0	10.0
100.5	10.4	10.4
101	10.7	10.7
110	11.7	11.7
125	12.0	12.0
150	12.3	12.3
200	12.5	12.5

For 0.1 mol • L^{-1} acetic acid and 0.1 mol • L^{-1} sodium hydroxide, it is evident from the titration curve that neither methyl orange nor methyl red can be used as indicators. The equivalence point is at pH 8.7, and it is necessary to use an indicator with a pH range on the slightly alkaline side, such as phenolphthalein, thymolphthalein, or thymol blue (pH range, as a base, 8.0-9.6). For the acid, with $K_a=10^{-7}$, the equivalence point is at pH=10, but the pH change rate in the stoichiometric point neighborhood is much less pronounced, owing to considerable hydrolysis. Phenolphthalein will change color after about 92 mL of alkali have been added. This change will occur to the equivalence point; thus, the endpoint will not

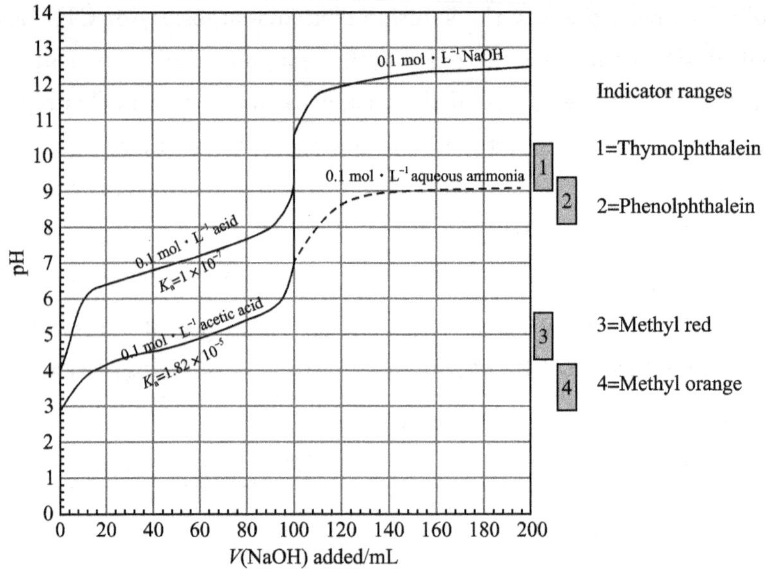

Figure 4.5 Neutralization curves of 100 mL of 0.1 mol · L^{-1} acetic acid ($K_a=1.82\times10^{-5}$) and of 0.1 mol · L^{-1} acid ($K_a=1\times10^{-7}$) with 0.1 mol · L^{-1} sodium hydroxide (calculated)

be sharp, and the titration error will be palpable. With thymolphthalein, however, the color change covers the pH range of 9.3-10.5; this indicator may be used. The endpoint will be sharper than for phenolphthalein but somewhat gradual, and the titration error will be about 0.2%. Acids that have dissociation constants less than cannot be satisfactorily titrated in a 0.1 mol · L^{-1} solution with a simple indicator.

In general, it may be stated that weak acids ($K_a>5\times10^{-6}$) should be titrated with phenolphthalein, thymolphthalein, or thymol blue as indicators.

4.6 Neutralization of a Weak Base with a Strong Acid

This may be illustrated by the titration of 100 mL of 0.1 mol · L^{-1} aqueous ammonia ($K_b=0.85\times10^{-5}$) with 0.1 mol · L^{-1} hydrochloric acid at the ordinary laboratory temperature. The equation gives the pH of the solution at the equivalence point:

$$\mathrm{pH}=\frac{1}{2}\mathrm{p}K_w-\frac{1}{2}\mathrm{p}K_b+\frac{1}{2}\mathrm{p}c=7-2.37+\frac{1}{2}(1.3)=5.28$$

For other concentrations, the pH may be calculated with sufficient accuracy as follows (compare the previous section):

$$[NH_4^+]\times[OH^-]/[NH_3]=K_b$$

$$\text{or } [OH^-]=[NH_3]\times K_b/[NH_4^+]$$

$$\text{or } \mathrm{pOH}=\lg([\mathrm{Salt}]/[\mathrm{Base}])+\mathrm{p}K_b$$

$$\text{or } \mathrm{pH}=\mathrm{p}K_w-\mathrm{p}K_b-\lg([\mathrm{Salt}]/[\mathrm{Base}])$$

After the equivalence point has been reached, the solution contains an excess of H^+ ions, hydrolysis of the salt is suppressed, and the subsequent pH changes may be assumed, with sufficient accuracy, to be those due to the excess of acid present.

The results computed above are represented graphically in Figure 4.6; the results for the titration of 100 mL of a 0.1 mol • L^{-1} solution of a weaker base ($K_b = 1 \times 10^{-7}$) is also included.

Neither thymolphthalein nor phenolphthalein can be used to titrate 0.1 mol • L^{-1} aqueous ammonia. The equivalence point is at pH 5.3, and it is necessary to use an indicator with a pH range on the slightly acid side (3-6.5), such as methyl orange, methyl red, bromophenol blue, or bromocresol green. The last-named indicators may be utilized to titrate all weak bases ($K_b > 5 \times 10^{-6}$) with strong acids.

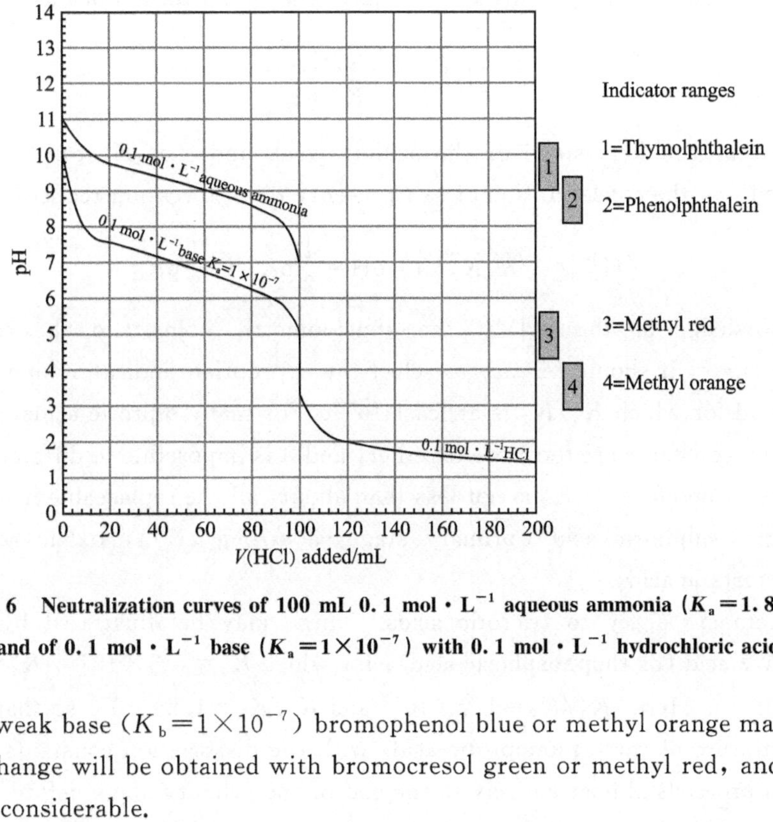

Figure 4.6 Neutralization curves of 100 mL 0.1 mol • L^{-1} aqueous ammonia ($K_a = 1.8 \times 10^{-5}$) and of 0.1 mol • L^{-1} base ($K_a = 1 \times 10^{-7}$) with 0.1 mol • L^{-1} hydrochloric acid

For the weak base ($K_b = 1 \times 10^{-7}$) bromophenol blue or methyl orange may be used; no sharp color change will be obtained with bromocresol green or methyl red, and the titration error will be considerable.

4.7 Neutralization of a Polyprotic Acid with a Strong Base

The shape of the titration curve will depend upon the relative magnitudes of the various dissociation constants. It is assumed that titrations occur in solutions of 0.1 mol • L^{-1} or stronger concentrations at the ordinary laboratory temperature. For a diprotic acid, if the difference between the primary and secondary dissociation constants is huge ($K_1/K_2 > 10^4$), the solution behaves like a mixture of two acids with constants K_1 and K_2, respectively; the considerations given previously may be applied. Thus, for sulfurous acid, $K_1 = 1.7\times10^{-2}$ and $K_2 = 1.0\times10^{-7}$ there will be a sharp change of pH near the first equivalence point, but for the second stage, the difference will be less pronounced yet just sufficient for the use of, such as thymolphthalein as the indicator (Figure 4.5). For carbonic acid, however, for which $K_1 = 4.3\times10^{-7}$ and $K_2 = 5.6\times10^{-11}$, only the first stage will be just discernible in the neutralization curve (Figure 4.5). The second stage is far too weak to exhibit any inflection point, and no suitable indicator is available for direct titration. As an indicator for the primary stage, thymol blue may be used, although a mixed indicator of thymol blue (3 parts) and cresol red (1 part) is more satisfactory; with phenolphthalein, the color change will be somewhat gradual, and the titration error may be several percent.

It can be shown that the pH at the first equivalence point for a diprotic acid is given by:

$$[H^+] = \sqrt{\frac{K_1K_2c}{K_1+c}}$$

Provided that the first stage of the acid is weak and that K_1 can be neglected by comparison with c, the concentration of salt present, this expression reduces to:

$$[H^+] = \sqrt{K_1K_2},\ \text{or pH} = \frac{1}{2}pK_1 + \frac{1}{2}pK_2$$

With knowledge of the pH at the stoichiometric point and the course of the neutralization curve, it should be easy to select the appropriate indicator for the titration of any diprotic acid for which K_1/K_2 is at least 10^{-4}. For many diprotic acids, however, the two dissociation constants are too close together, and it is impossible to differentiate between the two stages. Therefore, if K_2 is not less than about, all the replaceable hydrogen may be titrated, e. g., sulphuric acid (primary stage—a strong acid), oxalic acid, malonic, succinic, and tartaric acids.

Similar remarks apply to tricrotic acids. These may be illustrated by reference to phosphoric (V) acid (or thophosphoric acid), for which $K_1 = 7.5\times10^{-3}$, $K_2 = 6.2\times10^{-8}$, and $K_3 = 5\times10^{-13}$. Here, $K_1/K_2 = 1.2\times10^{-5}$ and $K_2/K_3 = 1.2\times10^{-5}$ so that the acid will behave as a mixture of three monoprotic acids with the dissociation constants given above. Neutralization proceeds almost entirely to the end of the primary stage before the secondary stage is appreciably affected. The second stage proceeds almost to completion before the tertiary stage is apparent. The pH at the first equivalence point is given approximately by $(1/2pK_1 + 1/2pK_2) = 4.6$, and at the second equivalence point by $(1/2pK_2 + 1/2pK_3) = 9.7$; in the very weak third stage, the curve is very flat, and no indicator is available for direct titration. The third equivalence point may be calculated approximately from the equation:

$$pH = \frac{1}{2}pK_w + \frac{1}{2}pK_a - \frac{1}{2}pc = 7 + 6.15 - \frac{1}{2}(1.6) = 12.35 \text{ for } 0.1\ mol\cdot L^{-1}\ H_3PO_4$$

For the primary stage (phosphoric(V) acid as a monoprotic acid), methyl orange, bromocresol green, or Congo red may be used as indicators. The secondary phosphoric(V) acid stage is very weak (see acid $K_a = 1 \times 10^{-7}$ in Figure 4.5), and the only suitable simple indicator is thymolphthalein; with phenolphthalein, the error may be several percent. A mixed indicator composed of phenolphthalein (3 parts) and 1-naphtholphthalein (1 part) is very satisfactory for the determination of the endpoint of phosphoric (V) acid as a diprotic acid. The experimental neutralization curve of 50 mL of 0.1 mol · L^{-1} phosphoric (V) acid with 0.1 mol · L^{-1} potassium hydroxide, determined by potentiometric titration, is shown in Figure 4.7.

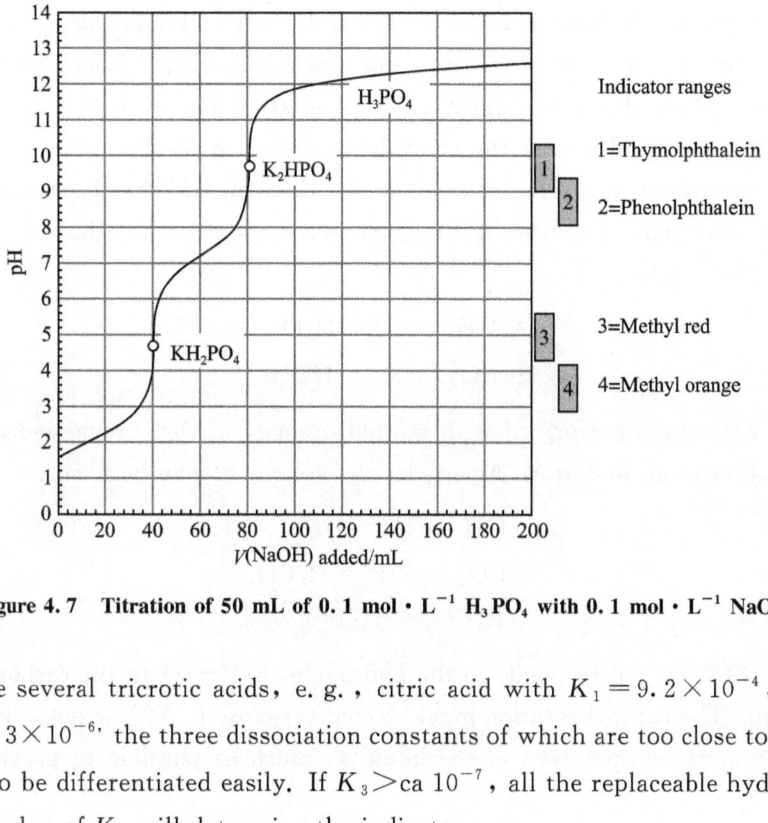

Figure 4.7 Titration of 50 mL of 0.1 mol · L^{-1} H_3PO_4 with 0.1 mol · L^{-1} NaOH

There are several tricrotic acids, e. g., citric acid with $K_1 = 9.2 \times 10^{-4}$, $K_2 = 2.7 \times 10^{-5}$, $K_3 = 1.3 \times 10^{-6}$, the three dissociation constants of which are too close together for the three stages to be differentiated easily. If $K_3 >$ ca 10^{-7}, all the replaceable hydrogen may be titrated; the value of K_3 will determine the indicator.

4.8 Determination of a Mixture of Carbonate and Hydroxide

Discussion. Two methods may be used for this analysis. In the first method, the total alkali (carbonate + hydroxide) is determined by titration with standard acid, using methyl orange, methyl orange-indigo carmine, or bromophenol blue as an indicator. The carbonate is precipitated with a slight excess of barium chloride solution in the second portion of the solution. Then, without filtering, the solution is titrated with standard acid using thymol blue or phenolphthalein as an indicator. The latter titration gives the hydroxide content, and by subtracting this from the first titration, the volume of acid required for the carbonate is obtained.

$$Na_2CO_3 + BaCl_2 \xlongequal{} BaCO_3(\text{insoluble}) + 2NaCl$$

The second method utilizes two indicators. It is the pH of half-neutralized sodium carbonate, i. e. , at the sodium hydrogen carbonate stage, is about 8. 3, but the pH changes comparatively slowly about the equivalence point; consequently, the indicator color change with phenolphthalein (pH range 8. 3-10. 0) or thymol blue [pH range (base) 8. 0-9. 6] is not too sharp. This difficulty may be overcome using a comparison solution containing sodium hydrogen carbonate of approximately the same concentration as the unknown and the same volume as an indicator. A simpler method is to employ a mixed indicator composed of 6 parts of thymol blue and 1 part of cresol red; this mixture is violet at pH 8. 4, blue at pH 8. 3, and rose at pH 8. 2. With this mixed indicator, the mixture has a violet color in an alkaline solution and changes to blue near the equivalence point; in the titration, the acid is added slowly until the solution assumes a rose color. All the hydroxide has been neutralized at this stage, and the carbonate is converted into hydrogen carbonate. Let the volume of standard acid consumed be v mL.

$$OH^- + H^+ = H_2O$$
$$CO_3^{2-} + H^+ = HCO_3^-$$

Another titration is performed with methyl orange, methyl orange-indigo carmine, or bromophenol blue as an indicator. Again, let the volume of acid be V mL.

$$OH^- + H^+ = H_2O$$
$$CO_3^{2-} + 2H^+ = H_2CO_3$$
$$H_2CO_3 = H_2O + CO_2$$

Then $V-2(V-v)$ corresponds to the hydroxide, $2(V-v)$ to the carbonate, and V to the total alkali. The titrated solution must be cold (as close to 0 ℃ as possible), and the tip of the burette must be immersed in the liquid as much as possible to prevent the loss of carbon dioxide.

Procedure A. Weigh out accurately about 2. 5 g of commercial sodium hydroxide (e. g. , in flake form) in a glass-stoppered weighing bottle. Transfer quantitatively to a 500 mL graduated flask and make up to the mark. Shake the flask well. Titrate 25 or 50 mL of this solution with standard 0. 1 mol · L^{-1} hydrochloric acid using methyl orange or methyl orange-indigo carmine as an indicator. Carry out two or three titrations: these should not differ by more than 0. 1 mL. This gives the total alkalinity (hydroxide + carbonate). Warm another 25 mL or 50 mL portion of the solution to 70 ℃ and slowly add 1% barium chloride solution from a burette or pipette in slight excess, i. e. , until no further precipitate is produced. Cool to room temperature, add a few drops of phenolphthalein to the solution, and titrate very slowly and with constant stirring with standard 0. 1 mol · L^{-1} hydrochloric acid; the endpoint is reached when the color changes from pink to colorless. If thymol blue is used as an indicator, the color changes from blue to yellow. The amount of acid used corresponds to the hydroxide present.

This method yields only approximate results because of the precipitation of basic barium carbonate in the presence of hydroxide. More accurate results are obtained by considering the above titration as preliminary to ascertain the approximate hydroxide content and then carrying out another titration. First, treat 25-50 mL of the solution with sufficient standard hydrochloric acid to neutralize most hydroxide, then heat and precipitate. Under these conditions, practically pure barium carbonate is precipitated.

$$1\ \text{mL}\ 1\ \text{mol} \cdot \text{L}^{-1}\ \text{HCl} = 0.040\ 1\ \text{g NaOH}$$

$$1\ \text{mL}\ 1\ \text{mol} \cdot \text{L}^{-1}\ \text{HCl} = 0.053\ 00\ \text{g}\ Na_2CO_3$$

Procedure B. The experimental details for the preparation of the initial solution are like those given under Procedure A. Titrate 25 mL or 50 mL of the cold solution with standard 0.1 mol • L^{-1} hydrochloric acid and methyl orange, methyl orange-indigo carmine, or bromophenol blue as indicators. Next, titrate another 25 mL or 50 mL of the cold solution, diluted with an equal volume of water, slowly with the standard acid using phenolphthalein or the thymol blue-cresol red mixed indicator; in the latter case, the color at the endpoint is rose.

Calculate the result as described in the Discussion above.

4.9 Determination of a Mixture of Carbonate and Hydrogen Carbonate

The two methods for this determination are modifications described in Section 4.8 for hydroxide/carbonate mixtures. In the first procedure, which is particularly valuable when the sample contains relatively large amounts of carbonate and small amounts of hydrogen carbonate, the total alkali is first determined in one portion of the solution by titration with standard 0.1 mol • L^{-1} hydrochloric acid using methyl orange, methyl orange-indigo carmine, or bromophenol blue as an indicator:

$$CO_3^{2-} + 2H^+ = H_2CO_3$$

$$HCO_3^- + H^+ = H_2CO_3$$

$$H_2CO_3 = H_2O + CO_2$$

Let this volume correspond to V mL 1 mol • L^{-1} HCl. In another sample, a measured excess of standard 0.1 mol • L^{-1} sodium hydroxide (free from carbonate) over that required to transform the hydrogen carbonate to carbonate is added:

$$HCO_3^- + OH^- = CO_3^{2-} + H_2O$$

A slight excess of 10 percent barium chloride solution is added to the hot solution to precipitate the carbonate as barium carbonate, and the excess of sodium hydroxide solution is immediately determined, without filtering off the precipitate, by titration with the same standard acid; phenolphthalein or thymol blue is used as the indicator. For example, if the volume of excess sodium hydroxide solution added corresponds to v mL of 1 mol • L^{-1} sodium hydroxide, and v' mL 1 mol • L^{-1} acid corresponds to the latter's excess, then $v-v'=$ hydrogen carbonate, and $V-(v-v')=$carbonate.

In the second procedure, a portion of the cold solution is slowly titrated with standard 0.1 mol • L^{-1} hydrochloric acid, using phenolphthalein or the thymol blue-cresol red mixed indicator. This (y mL) corresponds to half the carbonate (compare Section 4.8):

$$CO_3^{2-}+H^+=HCO_3^-$$

Another sample of equal volume is titrated with the same standard acid using methyl orange, methyl orange-indigo carmine, or bromophenol blue as an indicator. The volume of acid used (y mL) corresponds to carbonate + hydrogen carbonate. Hence, $2Y=$carbonate, and $y-2Y=$hydrogen carbonate.

Key Glossary

Amphoteric substance: A substance can react with both acids and bases.

Acid-base indicator: The indicator used for titration is a weak organic acid or weak organic base, which presents different colors in different pH solutions.

Discoloration range: The acid-base indicator changes from one color to another by visual inspection in a specific field of pH changes.

Titration curve: The curve indicating the change in the pH value of the solution with the titration process is called the titration curve.

Jump range: The solution parameters will change rapidly within ± 0.1% before and after the stoichiometric point. The sudden change of such parameters (such as pH in acid-base titration) is a titration jump.

Stepwise titration: For a polybasic acid, a titration method in which each endpoint is titrated successively.

Essential Exercises

Review Questions

1. Please briefly describe the relationship between K_a and K_b of the conjugate acid-base pair.

2. How does acidity affect the distribution of various weak acids?

3. What is the action principle, color change point, and acid-base indicator range?

4. Please briefly describe the similarities and differences of three different titration curves.

Cumulative Problems

1. The titration curve of HCl with the same concentration of 20.00 mL 0.1 mol • L^{-1} NaOH solution is as follows:

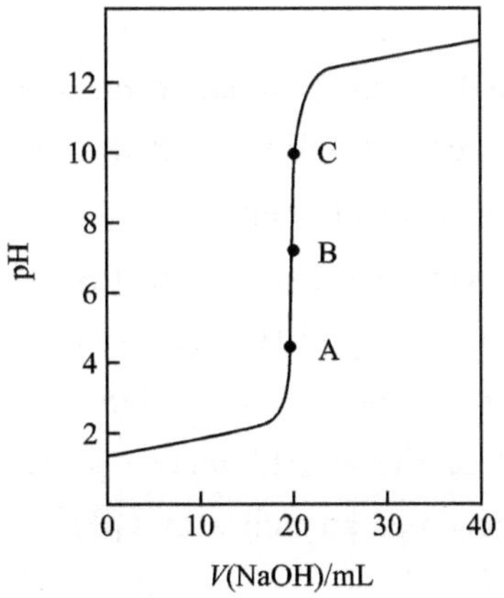

Please write out the composition of A, B, and C points in the curve and the formula for calculating $[H^+]$.

Point	The formula for calculating $[H^+]$	$V(NaOH)/mL$	Composition of the system
A		19.98	
B		20.00	
C		20.02	

2. The titration curve of HAc with the same concentration of 20.00 mL 0.1 mol · L^{-1} NaOH solution is as follows:

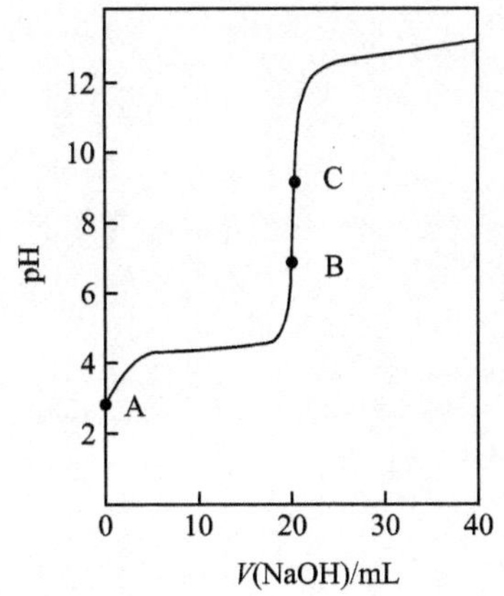

Please write out the composition of A, B, and C points in the curve and the formula for calculating $[H^+]$.

Point	The formula for calculating $[H^+]$	$V(NaOH)/mL$	Composition of the system
A		19.98	
B		20.00	
C		20.02	

3. The same formic acid (HA) concentration was titrated with 0.10 mol • L^{-1} NaOH. Please calculate the pH of the stoichiometric point and the endpoint error when methyl red was chosen as an indicator (pH 6.2 at the endpoint). [pK_a(HA)=3.74]

4. Calculate the pH of the stoichiometric point by titrating HA (pK_a=7.0) with 0.10 mol • L^{-1} NaOH at the same concentration. Please calculate the endpoint error if phenolphthalein is used as an indicator (pH is 9.0 at the endpoint).

5. CH_3NH_2 solution of the same concentration was titrated with 0.10 mol • L^{-1} HCl. Please calculate the pH of the stoichiometric point. Please calculate the endpoint error if methyl orange was chosen as an indicator (pH was 4.0 at the endpoint). [pK_b (CH_3NH_2)=3.38]

6. A sample contains Na_2CO_3 and $NaHCO_3$. First, weigh 0.301 0 g, using phenolphthalein as an indicator, 20.10 mL 0.106 0 mol • L^{-1} HCl was used during titration, methyl orange was used as an indicator, and 47.20 mL of HCl was used. Next, calculate the content of Na_2CO_3 and $NaHCO_3$ in the sample.

7. Calculate the pH of the following solutions:

(1)0.050 0 mol • L^{-1} Na_3PO_4 solution.

(2)1.00×10^{-4} mol • L^{-1} NH_4Cl solution.

(3)0.10 mol • L^{-1} NaCN solution.

Chapter 5

Precipitation Titration and Gravimetric Analysis

Key Points

- *Introduction: Precipitation Titration*
- *The Endpoint of a Method Based on Silver: The Mohr Method, The Volhard Method, The Fajans Method*
- *Gravimetric Analysis*

5.1 Precipitation Titration

5.1.1 Introduction

Precipitation titration is similar to complexometric and acid-base titration. It is also essential in chemical analysis. A **precipitation titration** method is one in which the reaction of a titrant with a sample leads to an insoluble precipitate. Some halogens and metal ions can be determined by using precipitation titration. The following situations must be satisfied in the process of precipitation titration:

(1) The rate of reaction must be quick.

(2) The reaction must be quantitative.

(3) The solubility of the precipitate is smaller.

(4) We can find a proper precipitant to determine the endpoint.

(5) The effect of selective precipitation of ions is small.

So, we also use silver ions as a precipitating agent to analyze the halogens or other anions in precipitation titration. This method is called **argentometry**. The expression of argentometry is:

$$Ag^+(aq) + X^-(aq) \rightleftharpoons AgX(s) \quad (X^- \text{ denotes halogens, } CN^-, SCN^- \text{ and so on})$$

This method can determine pesticides and Cl^- in the soil. Argentometry will be discussed in this chapter.

5.1.2 Principle of Argentometry

The principle of argentometry will be illustrated by using an experimental titration curve. The titration curve shows how the pH changes when the titrant is added in Chapter 4. Precipitation titration is different from acid-base titrations. The titration curve of precipitation titration denotes how the concentration of metal ions or anions changes when the titrant is added. We use the negative logarithm, pM, or pX. Example 5. 1 illustrates how to construct a titration curve.

Example 5. 1

20. 00 mL of 0. 100 0 mol · L^{-1} NaCl is treated with 0. 100 0 mol · L^{-1} $AgNO_3$. Construct a titration curve and analyze the process.

Solution:

In the first step, we must write the equilibrium expression of AgCl.

$$Ag^+(aq) + Cl^-(aq) \rightleftharpoons AgCl(s) \quad K_{sp}^{\ominus}=1.8\times10^{-10}$$

The expression for the solubility product is

$$K_{sp}^{\ominus}=[Ag^+][Cl^-]$$

Thus

$$-\lg K_{sp}^{\ominus}=-\lg([Ag^+][Cl^-])$$
$$pK_{sp}^{\ominus}=pAg+pCl$$

In this process, four regions of the titration curve require different kinds of calculations:

(1) $AgNO_3$ is not added to NaCl.

(2) Before the equivalence point, pAg or pCl is determined by excess Cl^-.

(3) At the equivalence point, pAg=pCl =1/2 $pK_{sp}^{\ominus}$=4. 9.

(4) After the equivalence point, pAg or pCl is determined by excess Ag^+.

We do the calculation process for each region as follows. Calculation results are shown in Table 5. 1. The titration curve is illustrated in Figure 5. 1.

(1) $AgNO_3$ is not added to NaCl:

$$[Cl^-]=0.1000\ mol\cdot L^{-1},\ pCl=1.0$$

(2) Before the equivalence point:

When 18. 00 mL of $AgNO_3$ have been added, the Cl^- is excess because $V(NaCl)$= 20. 00 mL. The concentration of Cl^- is:

$$[Cl^-]=\frac{0.1000\times2.00}{20.00+18.00}=5.26\times10^{-3}$$

And

$$pCl=2.3$$
$$[Ag^+][Cl^-]=K_{sp}^{\ominus}=1.8\times10^{-10}$$
$$pAg+pCl=-\lg K_{sp}^{\ominus}=9.74$$

So

$$pAg=7.4$$

The Cl^- is in excess when 19.98 mL of $AgNO_3$ has been added. Therefore, the calculation of the concentration of Cl^- is like 18.0 mL of $AgNO_3$.

(3) At the equivalence point:

The Ag^+ is sufficient to react with all Cl^- at the equivalence point. The pAg or pCl is determined by the solubility constant $K_{sp}^{\ominus}$.

$$pAg=pCl=1/2pK_{sp}^{\ominus}=4.9$$

(4) After the equivalence point:

When 20.02 mL of $AgNO_3$ has been added, the $AgNO_3$ solution is in excess. pAg or pCl is determined by excess Ag^+. The concentration of Ag^+ is:

$$[Ag^+]=\frac{0.1000\times 0.02}{20.00+20.02}=5.0\times 10^{-5}$$

And

$$pAg=4.30$$
$$[Ag^+][Cl^-]=K_{sp}^{\ominus}=1.8\times 10^{-10}$$
$$pAg+pCl=-\lg K_{sp}^{\ominus}=9.7$$

So

$$pAg=9.7-4.3=5.4$$

Table 5.1 Calculation of the titration curve for 20.00 mL of 0.1000 mol · L^{-1} $AgNO_3$ treated with 0.1000 mol · L^{-1} NaCl

$V(AgNO_3)$ added/mL	$[Cl^-]$	pCl	pAg
0	10^{-1}	1	—
18.00	$10^{-2.3}$	2.3	7.4
19.80	$10^{-3.3}$	3.3	6.5
19.98	$10^{-4.3}$	4.3	5.5
20.00	$10^{-4.9}$	4.9	4.9
20.02	$10^{-5.5}$	5.4	4.3
20.20	$10^{-6.5}$	6.5	3.3
22.00	$10^{-7.5}$	7.5	2.3
40.00	$10^{-8.3}$	8.3	1.5

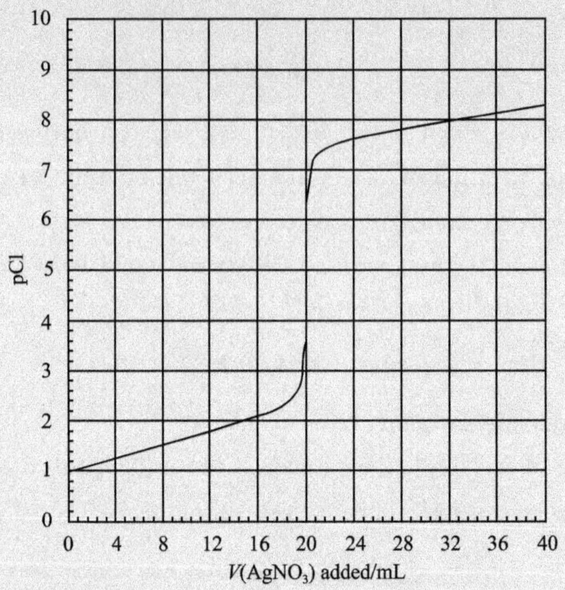

Figure 5. 1 Calculation of the titration curve

Are you wondering what are the effect factors of the titration curve?

We can analyze the Cl^- by using Argentometry and analyzing the Br^-, I^-. Figure 5. 2 shows that the titration curves of Cl^-, Br^-, I^- are compared.

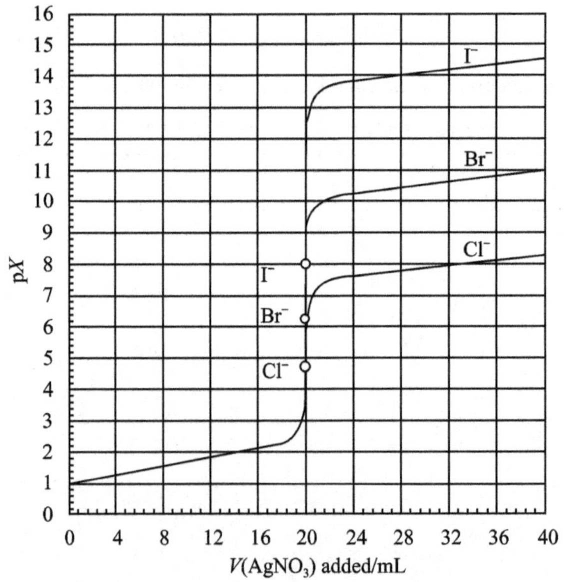

Figure 5. 2 The comparison of titration curves for the analysis of 0. 100 0 mol · L^{-1} $AgNO_3$ by its titration with 20 mL of 0. 100 0 mol · L^{-1} NaCl, NaBr, NaI

The titration curves are like acid-base curves or complexometric curves. The effect factors of the titration curve are $K_{sp}^{\ominus}$ and the concentration of titrant or standard solution. In Figure 5.2, the value of AgI ($K_{sp}^{\ominus}=8.3\times10^{-17}$) is smaller than AgBr ($K_{sp}^{\ominus}=5.0\times10^{-13}$) and AgCl ($K_{sp}^{\ominus}=1.8\times10^{-10}$). In other words, the solubility of AgI is also smaller. The range of pI is wide.

5.1.3 Determining the Endpoint

1. The Mohr Method

This method is mainly applied to the titration of chloride and bromide ions with $AgNO_3$ solution. The indicator is the K_2CrO_4 solution.

(1) The principle of the Mohr Method is shown as follows. Figure 5.3 illustrates that $AgNO_3$ is added to the solution containing Cl^- and CrO_4^{2-}. The AgCl is precipitated first because the solubility of AgCl is smaller. The color of AgCl is white. The Ag_2CrO_4 begins to precipitate when the concentration of Cl^- becomes less at the equivalence point.

So, the endpoint is signaled by the appearance of red silver chromate (Ag_2CrO_4).

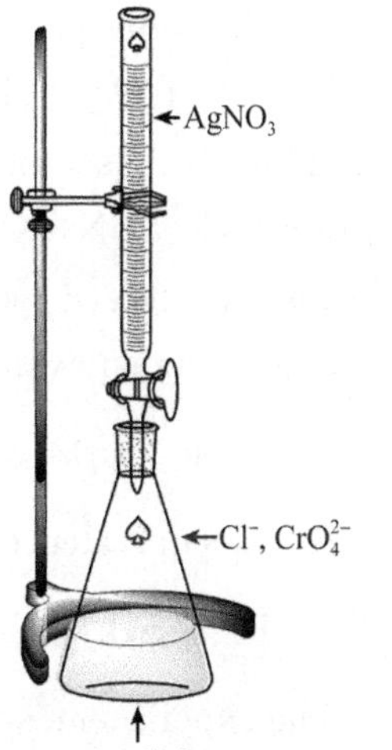

Figure 5.3 The illustration of the Mohr Method

The expressions of equilibrium are:

Before the equivalence point, $Ag^+(aq)+Cl^-(aq)\rightleftharpoons AgCl(s)$ $\quad K_{sp}^{\ominus}=1.8\times10^{-10}$

At the equivalence point, $2Ag^+(aq)+CrO_4^{2-}(aq)\rightleftharpoons Ag_2CrO_4(s)$ $\quad K_{sp}^{\ominus}=1.2\times10^{-12}$

$$K_{sp}^{\ominus}=[Ag^+][Cl^-]$$

$$[Ag^+][Cl^-]=1.8\times10^{-10}$$

$$K_{sp}^{\ominus}=[Ag^+][Cl^-],[Ag^+]=[Cl^-]=\sqrt{K_{sp}^{\ominus}}=1.34\times10^{-5}(\text{mol}\cdot\text{L}^{-1})$$

$$K_{sp}^{\ominus}=[Ag^+]^2[CrO_4^{2-}]$$

Thus

$$[CrO_4^{2-}]=\frac{K_{sp}^{\ominus}}{[Ag^+]^2}=\frac{1.2\times10^{-12}}{(1.34\times10^{-5})^2}=6.68\times10^{-3}(\text{mol}\cdot\text{L}^{-1})$$

In the endpoint, the concentration of CrO_4^{2-} is 6.68×10^{-3} ($\text{mol}\cdot\text{L}^{-1}$).

(2) The effect factor of the Mohr Method. The concentration of CrO_4^{2-}. In the endpoint, the concentration of CrO_4^{2-} is $6.68\times10^{-3}\ \text{mol}\cdot\text{L}^{-1}$. We usually add 1-2 mL of 5% $K_2Cr_2O_4$ in 100 mL solution and add excess $AgNO_3$. Because the yellow color of CrO_4^{2-} can affect the determination of the endpoint.

Considering the value of pH. The titration is performed in neutral or slightly basic medium pH=6.5-10.5. The reason is that silver hydroxide will be formed at pH>10.5 or chromic acid at pH<6.5. The sodium hydrogen carbonate or borax can be used in an acid medium with pH< 6.5.

2. The Volhard Method

There are two ways to use the Volhard method. One is a direct method to detect Ag^+. The other is an indirect method to detect halides. The indicator is $NH_4Fe(SO_4)_2 \cdot 12H_2O$ solution.

(1) The principle of the Volhard Method is shown as follows.

To detect Ag^+ (Figure 5.4): $NH_4Fe(SO_4)_2$ a solution containing Fe^{3+} is used as an indicator. A standard thiocyanate solution (NH_4SCN, KSCN, or NaSCN) is used as a titrant. The appearance of the red complex signals the endpoint, $[Fe(SCN)]^{2+}$.

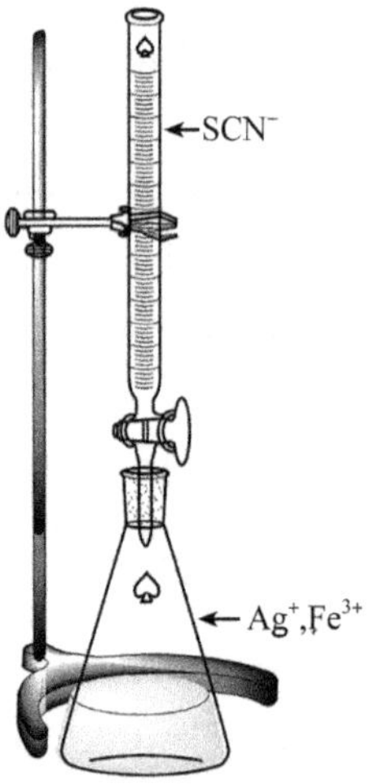

Figure 5.4 The illustration of the Volhard Method

Before the equivalence point,

$$Ag^+(aq) + SCN^-(aq) \rightleftharpoons AgSCN\downarrow \text{(white)}$$

At the equivalence point,

$$Fe^{3+}(aq) + SCN^-(aq) \rightleftharpoons [Fe(SCN)]^{2-}\downarrow \text{(red)}$$

The experiment shows that the red $[Fe(SCN)]^{2+}$ complex can be detected at 6.0×10^{-6} mol·L^{-1}.

An indirect method to detect halides: the excess $AgNO_3$ standard solution is first added to the Cl^- ion solution. The excess of Ag^+ is determined by back-titration with a standard thiocyanate solution. The expressions are:

$$Ag^+ + Cl^-(aq) \rightleftharpoons AgCl\downarrow + Ag^+_{excess} \quad \text{Titration reaction}$$

$$Ag^+_{excess} + SCN^-(aq) \rightleftharpoons Ag(SCN) \quad \text{Back titration reaction}$$

$$Fe^{3+}(aq) + SCN^-(aq) \rightleftharpoons [Fe(SCN)]^{2+}\downarrow \text{(red)} \quad \text{Endpoint reaction}$$

The sum of expressions is:

$$[Fe(SCN)]^{2+} + AgCl \rightleftharpoons Fe^{3+}(aq) + AgSCN + Cl^-$$

(2) The titration conditions of the Volhard Method. Considering pH. A strong acid medium is required because Fe^{3+} will react with OH^- in a neutral or basic medium.

The concentration of indicator. When the concentration of Fe^{3+} is 0.015 mol·L^{-1}, the endpoint can be detected perfectly, and the end error is smaller than 0.1%.

The effect of complex ions or strong oxidants (such as Hg^{2+}). We must remove these ions from the solution before we analyze halide ions.

The effect of precipitation transformation is an indirect method. A significant error will occur. There are two ways to solve this error. One is the removal of AgCl by filtration. The other is adding nitrobenzene and shaking vigorously.

When we analyze I^-, the indicator should be added after adding the excess $AgNO_3$ solution.

In a word, we cannot only analyze the halide ions but also analyze the other complex ions, such as PO_4^{3-}, AsO_4^{3-} by using the Volhard Method.

3. The Fajans Method

The Fajans Method, named after Kazimierz Fajans, is also called **the adsorption indicator method.** This method is applied to analyze halide ions and thiocyanate ions. The adsorption indicator determines the endpoint. The adsorption indicator is mainly dye, such as dichlorofluorescein, fluorescein, eosin, etc.

(1)The principle of the Fajans Method. The stoichiometric titration reaction is:

$$Ag^+(aq) + Cl^-(aq) \rightleftharpoons AgCl$$

$$\underset{}{HFIn} \rightleftharpoons H^+(aq) + \underset{\text{(yellow)}}{FIn^-}$$

$$AgCl + Ag^+ + \underset{\text{(yellow)}}{FIn^-} \rightleftharpoons \underset{\text{(pink)}}{AgCl \cdot Ag^+ \vdots FIn^-}$$

The HFIn is the symbol of fluorescein. The endpoint is marked by the yellow suspension turning pink. Before the endpoint of the titration, chloride ions remain in excess. They adsorb on the AgCl surface, imparting a negative charge to the particles. After the equivalence point, excess silver (I) ions adsorb on the AgCl surface, imparting a positive charge. Anionic dyes such as dichlorofluorescein and fluorescein are attracted to the particles and undergo a color change upon adsorption, representing the endpoint. Eosin (tetrabromofluorescein) is suitable for titrating against bromide, iodide, and thiocyanate anions, giving a sharper endpoint than dichlorofluorescein. It is unsuitable for titrating against chloride anions because it binds to AgCl more strongly than chloride.

(2) The titration conditions of the Fajans Method. Considering the value of pH. Table 5. 2 illustrates the different pH of the indicators compared.

Table 5. 2 The comparisons of different pH of the indicators

Indicators	$pK_a^\ominus$	Color changes	pH
Fluorescein	7. 0	Yellow→Pink	7-10
Dichlorofluorescein	4. 0	Yellow→Pink	4-10
Eosin	2. 0	Pink→Purple	2-10

The situation of titration is dark.

5. 2 Gravimetric Analysis

5. 2. 1 Introduction

Gravimetric analysis is a set of methods used in analytical chemistry to quantitatively determine the analyte (the ion being analyzed) based on its mass. The principle behind this type of analysis is that once an ion's mass has been determined as a unique compound, that known measurement can be used to determine the same analyte's mass in a mixture if the relative quantities of the other constituents are known. This method has four types: volatilization, precipitation, electro-analytical, and miscellaneous physical method. This method can be applied in agriculture. For example, the soil's dry matter and water content are determined by gravimetric analysis.

There are advantages and disadvantages to this method. The advantages are that the mass of a substance can be measured accurately, and the error is smaller. The disadvantages are that the time is longer, and few products are completely quantitative precipitation. So, the range of gravimetric applications is far from that of titration analysis.

5.2.2 General Process of Gravimetric Analysis

(1) The general process. The analysis procedure has several steps.

① The mass of a sample is measured.

② The sample is dissolved.

③ Choosing the proper situation. We can adjust pH to form the proper precipitate and remove some interference from some reagents.

④ The precipitating reagent is added. We should consider the concentration, heating ("digestion"), or pH. Digestion can help reduce the amount of coprecipitation.

⑤ The precipitate will be collected by filtering the solution.

⑥ Washing the precipitate.

⑦ Drying and weighing precipitates.

⑧ Calculation of the content of a sample.

For example, we want to know the content of $BaSO_4$ or CaO in a sample. The processes of gravimetric analysis are:

$$Ba^{2+} + SO_4^{2-} \rightarrow BaSO_4 \downarrow \xrightarrow{\text{filter, wash}} \xrightarrow{800\ ^\circ C\text{, combustion}} BaSO_4$$

$$Ca^{2+} + C_2O_4^{2-} \rightarrow CaC_2O_4 \cdot 2H_2O \downarrow \xrightarrow{\text{filter, wash}} \xrightarrow{\text{dry, combustion}} CaO$$

(2) The demands of precipitation form and weighing form in thegravimetric analysis method.

① The demands of precipitation form:

The solubility of the precipitate is small.

The precipitate is easy to filter and wash.

The precipitation form is easy to convert to the weighing form.

② The demands of weighing form:

The target product has stable physical and chemical properties.

The target product has larger molar solubility.

5.2.3 Solubility of Precipitate and Effect Factor of Solubility

(1) The solubility of the precipitate. We learned the equilibrium of precipitate. The

expression is

$$MA \rightleftharpoons M^{+}(aq) + A^{-}(aq)$$

The expression of the solubility product constant $K_{sp}^{\ominus}$ is

$$K_{sp}^{\ominus} = [M^{+}][A^{-}] = S^{2}$$

Solubility, S is the greatest amount of a substance that will dissolve in equilibrium. S is the symbol of complete precipitation. When the value of S is small, the substance can be precipitated completely.

(2) The effect factor of solubility. Solubility is significant in the gravimetric analysis. Several factors affect solubility.

①The effect of common-ion. The solubility will decrease when the common ion is added to the solution. Example 5. 2 illustrates the effect of common-ion.

Example 5. 2

SO_4^{2-} is determined by using gravimetric analysis. An equal amount of $BaCl_2$ and an excess amount of 0. 01 mol • L^{-1} $BaCl_2$ are respectively added to 200 mL of the solution, including SO_4^{2-}.

Calculate the mass of dissolved $BaSO_4$.

The dissolution equilibrium is:

$$BaSO_4(s) \rightleftharpoons Ba^{2+}(aq) + SO_4^{2-}(aq) \qquad K_{sp}^{\ominus} = 1.1 \times 10^{-10}$$

Solution:

(1) An equal amount of $BaCl_2$ is added to the solution.

$$S = \sqrt{K_{sp}^{\ominus}} = \sqrt{1.1 \times 10^{-10}} = 1 \times 10^{-5}(mol \cdot L^{-1})$$

The mass of dissolved $BaSO_4$ is

$$m = S \times 233.4 \times 0.2 \approx 0.000\,5\ (g) > 0.2(mg)$$

The error is 0. 2 mg. The mass of dissolved $BaSO_4$ is larger than the error. So, the SO_4^{2-} is not determined by using gravimetric analysis.

(2) Excess amount of 0. 01 mol • L^{-1} $BaCl_2$ is added to the solution.

$$[Ba^{2+}] = 0.01\ mol \cdot L^{-1}$$

$$S = \frac{K_{sp}^{\ominus}}{0.01} = 1.1 \times 10^{-8}(mol \cdot L^{-1})$$

The mass of dissolved $BaSO_4$ is

$$m = S \times 233.4 \times 0.2 \approx 5 \times 10^{-7}(g) < 0.2(mg)$$

The mass of dissolved $BaSO_4$ is smaller than the error. So, the SO_4^{2-} can be determined by using gravimetric analysis. The solubility will be smaller because of the common-ion.

②The effect of solubility of salts of bases. The molar solubility of this solution will be larger when a strong electrolyte (such as KNO_3) is added to the solution.

③The effect of the value of pH. The molar solubility of this solution will be larger when the pH value is changed.

④The effect of the complex. The molar solubility of this solution will be larger because of the complex effect.

⑤The effect of other factors, such as temperature, solvent polarity, colloid formation, and so on.

5.2.4 Precipitation Formation and Selection of Precipitation Condition

1. The precipitation formation

Figure 5.5 shows the process of precipitation formation. There are two main types of precipitation. One is crystal precipitation. The other is amorphous precipitation. The difference between the two types is mainly particle size in the crystallization process. What happens in the process of crystallization? The crystallization occurs in two phases: nucleation and particle growth. When the substance begins to precipitate, very tiny particles are formed. We call the small particles "**nuclei**", and the forming process is called "**nucleation**". Then, the molecules in the solution will form aggregates in three dimensions, and the particles become larger. We call this process "**particle growth**". In the crystallization process, the solute will increase, and we call the solution "supersaturated" more than that of equilibrium.

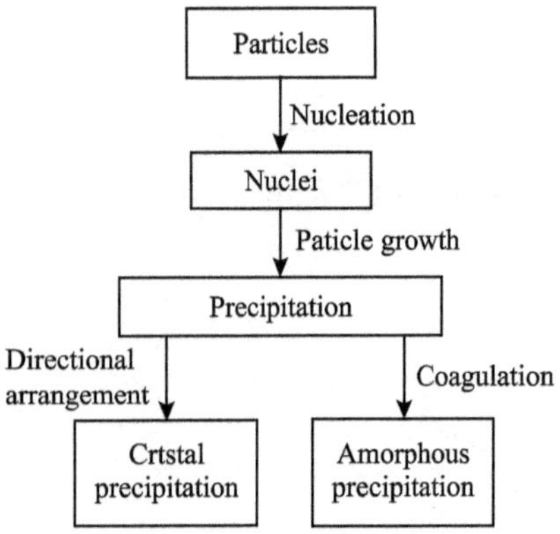

Figure 5.5 The illustration of precipitation formation

Nucleation is slower, and the nuclei have a chance to grow larger in a less supersaturated solution. Crystal precipitation will be formed in this solution. In contrast, in a highly supersaturated solution, nucleation is faster. Then, the suspensions of tiny particles or collide are formed.

So, in a highly supersaturated solution, amorphous precipitation is formed.

In the gravimetric analysis method, the ideal product should be crystal precipitation that is pure, easily filterable, and insoluble. So, we must consider supersaturation. In the early part of the twentieth century, Von Weimarn found that the particle size of precipitates was relative to supersaturation. The expression is:

$$R=\frac{Q-S}{S} \qquad \textbf{(Equation 5.1)}$$

Q is the molar concentration of mixed reagents before precipitation forms. S is the molar solubility of a precipitate. R is relative supersaturation.

When the relative supersaturation increases, the rate of precipitation will increase. So, the relative supersaturation should be slow to get perfect crystals.

2. Selection of precipitation condition

(1) Choosing a less supersaturated solution.

(2) The precipitant is slowly added to the solution under stirring conditions.

(3) Precipitation from a hot solution. The solubility of precipitates increases with temperature. Therefore, according to equation 5.1, the relative supersaturation will decrease with increased solubility.

(4) Digestion of precipitate. **Digestion, or precipitate aging**, happens when a freshly formed precipitate is left in the solution from which it precipitates, usually at a higher temperature. It results in cleaner and bigger particles. The physical-chemical process underlying digestion is called Ostwald ripening. In the formation process of crystal precipitation, the precipitate must be digested to give a perfect crystal. However, in the formation process of amorphous precipitation, digestion of precipitate should not occur. The reason is that colloidal precipitates may occur in high supersaturation. As a result, colloidal precipitates will be coagulated into large particles. We must heat the solution or add the electrolyte to solve this problem.

5.2.5 Purity of Precipitation

We always want to get pure precipitation in gravimetric analysis. However, the adsorption of other ions occurs at the crystal's surface. Coprecipitation is the main factor that affects the purity of precipitation. Coprecipitation is the carrying down by a precipitate of substances normally soluble under the conditions employed. Adsorption inclusion and occlusion are the three main mechanisms of coprecipitation. The surface of the precipitate will attract ions when the precipitate is formed. This process is called "adsorption". Adsorbed impurities can be removed by washing the precipitate. Inclusion occurs when the impurity occupies a lattice site in the crystal structure of the carrier, resulting in a crystallographic defect. Occlusion occurs when an adsorbed impurity gets physically trapped inside the crystal as it grows. If the impurities occur in occlusion, they cannot be removed by washing the precipitate.

5.2.6 Filtration, Washing, Drying, and Ignition of Precipitation

1. The filtration of precipitation

The precipitate must be removed from the mother liquor after the precipitate is completely formed. We can isolate the precipitate by using filter paper or a crucible.

The filter paper is the most common filtering medium. There are three types of filter paper: fast (the size of retained particles > 20-25 μm), medium (the size of retained particles >16 μm), and slow (the size of retained particles > 2.7 μm). Figure 5.6 shows the precipitate being isolated using filter paper.

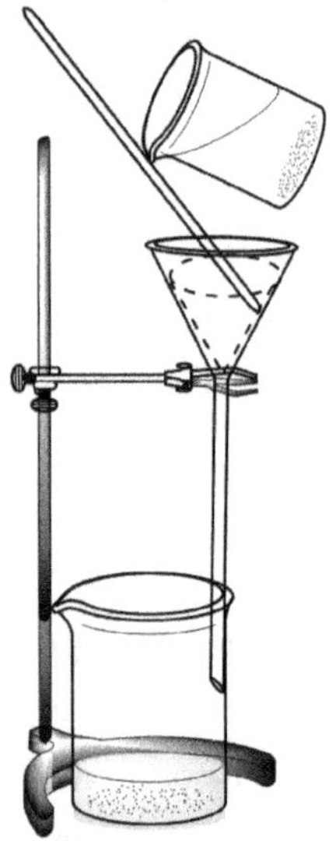

Figure 5.6 Use of the filter paper

In Figure 5.6, the filter paper is folded to form a cone and placed into a glass funnel. Then, the supernatant is transferred to the filter paper cone. Finally, the precipitate is transferred to the filter paper.

Another method for filtering the precipitate is a filtering crucible. There are three types: coarse (retaining particles > 40-60 μm), medium (retaining particles > 10-15 μm), and fine (retaining particles > 4-5.5 μm).

2. Washing precipitation (or rinsing precipitation)

After filtration, the precipitate should be washed to remove the impurities and residual solution. Choosing the proper rinsing medium is very important. We can choose some organic solvents, diluted precipitants, electrolytes (such as ammonium salts), etc. The electrolyte, especially a volatile inverted electrolyte, is added to the solution to prevent precipitation from peptization. The process by which the precipitate can revert to a small particle is called "peptization". The volatile inverted electrolyte is easy to remove by drying the precipitate.

3. Drying and ignition of precipitation

The remaining moisture is removed, and the precipitate is converted to a more chemically stable form through drying and ignition precipitation. The method of filtration and the precipitate's form should be considered when we choose the temperature and method of drying. If the precipitate is filtrated by filter paper, the filter paper and the precipitate are transferred to a platinum crucible, and the filter paper is charred at a low temperature. Then, the temperature slowly increases to a higher temperature. Don't worry if you see the smoke. We usually dry and ignite the precipitate by using a muffle furnace. If a filtering glass crucible filtrates the precipitate, it should be dried in an oven because the filtering glass crucibles cannot withstand high temperatures.

After drying and ignition of the precipitate, it is allowed to cool. Then, it is weighed, and the final mass can be calculated. The initial mass of the empty crucible is subtracted from the final mass of the crucible containing the sample. Since the composition of the precipitate is known, it is simple to calculate the mass of the analyte in the original sample.

5.2.7 Calculation of Gravimetric Analysis

We can calculate the mass of the final product by using Equation 5.2.

$$\text{Chemical factor} = \frac{m \times M_B}{n \times M_A} \qquad \textbf{(Equation 5.2)}$$

M_A is the sample's relative molecular mass; M_B is a determined substance's relative molecular mass; m and n are coefficients. For instance, calculating the mass of S from $BaSO_4$, m and n is 1. Calculating the mass of Cr_2O_3 from $BaSO_4$, m and n are 1 and 2. Table 5.3 shows the chemical factors of different substances. Example 5.3 illustrates the calculation of gravimetric analysis.

Table 5.3 Chemical factors of different substances

Analyte	Precipitate formed	Chemical factor
Ba	$BaSO_4$	$Ba/BaSO_4=0.5883$
S	$BaSO_4$	$S/BaSO_4=0.1374$
Fe	Fe_2O_3	$2Fe/Fe_2O_3=0.6995$
MgO	$Mg_2P_2O_7$	$2MgO/Mg_2P_2O_7=0.3622$
P	$Mg_2P_2O_7$	$2P/Mg_2P_2O_7=0.2783$
P_2O_3	$Mg_2P_2O_7$	$P_2O_5/Mg_2P_2O_7=0.6376$

Example 5.3

S of the pyrite is transformed to $BaSO_4$ by using gravimetric analysis. The mass of the sample is 0.2436 g. The precipitate, $BaSO_4$, weighed 0.5218 g. Calculate the percentage of the mass of S, $\omega(S)$.

Solution:

$$\text{Chemical factor}=\frac{m\times M_B}{n\times M_A}$$

and

$$m\times M_B=\text{Chemical factor}\times(n\times M_A) \qquad \textbf{(Equation 5.3)}$$

According to Table 5.3, we know:

$m=1$, $n=1$, chemical factor $S/BaSO_4=0.1374$

Thus

the mass of $S=0.5218\times0.1374=0.07170$ (g)

$$\omega(S)=\frac{0.07170}{0.2436}\times100\%=29.43\%$$

Key Glossary

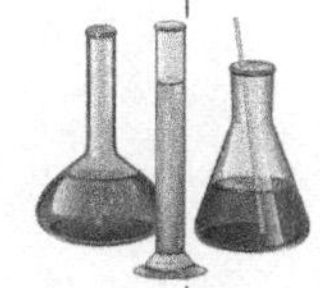

Precipitation titration is a titration method in which the reaction of a titrant with a sample leads to an insoluble precipitate. It includes the Mohr Method, the Volhard Method, and the Fajans Method.

Essential Exercises

Review Questions

1. Describe the principle and situations of precipitation titration.

2. Compare the Mohr Method, the Volhard Method, and the Fajans Method.

3. What is "gravimetric analysis"? What are some advantages of this method?

4. In your words, illustrate the application of precipitation titration and gravimetric analysis in agriculture.

Cumulative Problems

1. 1.579 3 g NaCl is dissolved in pure water, and this solution is prepared in the 250 mL volumetric flask. Then, 20 mL of this prepared solution is titrated with 19.20 mL of $AgNO_3$. Next, calculate the concentration of $AgNO_3$.

2. The application of the Mohr Method. 0.167 9 g of impure KCl is dissolved in pure water. This solution is titrated by 20.12 mL of 0.101 3 mol • L^{-1} $AgNO_3$. Predict the purity of the sample.

3. A sample of pesticide weighs 0.734 4 g. This sample is decomposed by the melting method, and the water is added to the decomposed solution. Then, the solution containing F^- is filtered. F^- is converted to PbClF by adding excess HCl and $Pb(NO_3)_2$. 45.00 mL of 0.300 0 mol • L^{-1} $AgNO_3$ and benzene are added to this solution after diluting HNO_3 is added to the PbClF precipitate. $NH_4Fe(SO_4)_2$ is used as an indicator. Excess Ag^+ requires 10.50 mL of 0.200 0 mol • L^{-1} NH_4SCN solution for titration. Calculate the percentage of the mass of Na_2SeF_6 and F.

4. Calculate the chemical factor.

(a) Fe from Fe_2O_3; (b) P_2O_5 from $Mg_2P_2O_7$; (c) Ba from $BaSO_4$.

5. Analyst the mass of Fe of $FeSO_4 \cdot 7H_2O$. Fe^{2+} is transformed to Fe^{3+}. The precipitate, $Fe(OH)_3$, occurs when an ammonia solution is used. $Fe(OH)_3$ is transformed into Fe_2O_3 by filtration, washing, and ignition. What is the mass of the sample if Fe_2O_3 weighs 0.2 g?

6. Calculate and explain the following phenomena.

(a) The solubility of CaF_2 at pH 3.00 is larger than at pH 4.00.

$[K^{\ominus}_{sp}(HF)=6.6\times10^{-4}, K^{\ominus}_{sp}(CaF_2)=2.7\times10^{-11}]$

(b) To analyze the mass of Al, we usually choose the $Al(C_9H_6ON)_3$ instead of Al_2O_3. Why? $\{M[Al(C_9H_6ON)_3]=460.00, M(Al)=26.98\}$

Chapter 6

Complexometric Titration

Key Points

- *Characteristics of EDTA-Metal Ion Complexes*
- *Side Reaction Coefficient and Conditional Formation Constant*
- *Metal Ion Indicator*
- *EDTA Titration Curve*
- *Conditions of Complexometric Titration*
- *Methods and Applications of Complexometric Titration*
- *Calculation of Complexometric Titration*

6.1 Overview

Complexometric titration is the titration analysis based on the reaction of complexes. For example, in the case of $AgNO_3$, titration of CN^-, Ag^+ reacts with CN^- to generate the complex ion. The equation is as follows:

$$Ag^+ + 2CN^- \rightleftharpoons [Ag(CN)_2]^-$$

When titration reaches the stoichiometric point, a slight overdose of Ag^+ may combine with $[Ag(CN)_2]^-$. The solution turns cloudy to form a white precipitation $Ag[Ag(CN)_2]$ and indicates the titration's end.

There must be some conditions for the complexometric titration:

(1) The complexation reaction must be complete. The formation constant of the complex (K_f) is large enough.

(2) The reaction should be carried out in a certain chemical equation, and the coordination ratio should be constant.

(3) The rate of the reaction is quick.

(4) A proper method could be used to detect the endpoint.

The complexation reactions are of great universality, but not all can satisfy the above conditions.

6.1.1 Inorganic Ligands and Simple Complexes

Many inorganic compounds could act as ligands for metal ions. What all of these ligands have in common is they have a pair of electrons that can be donated to a metal ion. For example, the ions: NH_3, CN^-, F^-, Cl^-,etc., can only donate one pair of electrons to a metal ion and are called a monodentate ligand or simple ligand, where monodentate means "one-toothed". These monodentate ligands coordinate the metal ions individually, often forming the ML_n-type simple complexes. For example, in the complexation reaction of Cu^{2+} and NH_3, the complex ions $[Cu(NH_3)]^{2+}$, $[Cu(NH_3)_2]^{2+}$, $[Cu(NH_3)_3]^{2+}$, $[Cu(NH_3)_4]^{2+}$ are stepwise generated. Therefore, in addition to the individual reaction (such as Ag^+ and CN^-, Hg^{2+} and Cl^-), inorganic ligands could rarely be used in complexation titration in analytical chemistry, but they are commonly used as a masking agent, auxiliary complexing agent, and color agent.

6.1.2 Organic Ligands and Chelate Complexes

Organic ligands often contain more than one donor atom. They could be divided into subcategories based on their binding sites for a metal ion. Ethylenediamine, $\ddot{N}H_2CH_2CH_2\ddot{N}H_2$, is an example of a bidentate ligand with two binding sites for a metal ion. Other organic ligands may have three, four, or even more sites for binding metals, called tridentate, tetradentate, or polydentate ligands. Most organic ligands, known as **chelating agents**, could produce a ring-type structure when they form a complex with a metal ion. This type of complex is called the **chelate complex**. The stability of the chelate complex is higher than that of simple complexes. Table 6.1 compares three complexes formed by Cu^{2+} and ammonia, ethylenediamine, and triethylenetetramine.

Table 6.1 Comparison of Cu^{2+} and ammonia, ethylenediamine, triethylenetetramine complexes

Complexes	Coordination ratio	Number of chelate ring	$\lg K_f$*
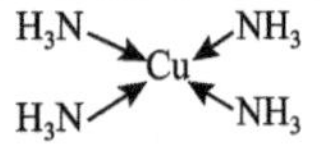	1∶4	0	12.6
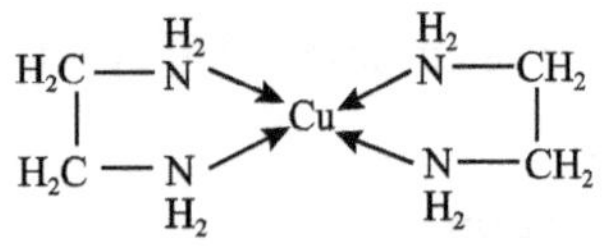	1∶2	2	19.6
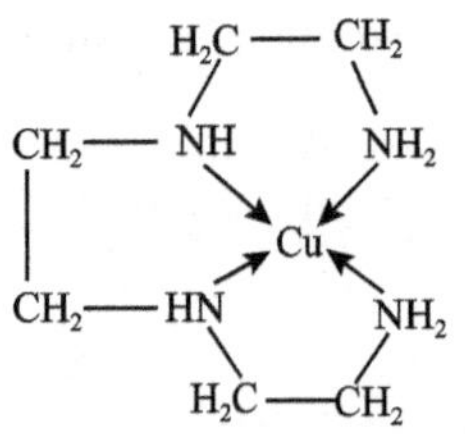	1∶1	3	20.6

* $\lg K_f$ is the formation constant of coordination compounds.

Since organic ligands contain multiple donor atoms, so hierarchical coordination could be reduced or eliminated. Furthermore, the disadvantages of inorganic ligands could be overcome, especially because of the high stability of the chelate generated from the organic ligands. Currently, ammonia-carboxyl ligands are the most used in complexometric titration.

The ammonia-carboxylation ligands contain the group of ammonia $\ddot{N}$ (with three single bonds) and carboxy oxygen —C(=O)—$\ddot{O}$— . The former is easily coordinated with Cu, Ni, Zn, Co, and Hg metal ions. The latter could coordinate with most metal ions. Therefore, the ammonia-carboxylation ligands could generate complex ions with all metal ions.

In a complexometric titration, the most common ammonia-carboxylation ligands are EDTA (ethylenediaminetetra-acetic acid), CyDTA (or DCTA, cyclohexane diamine tetraacetic acid), EDTP (ethylenediamine tetrapropionic acid), TTHA (triethyl tetramine hexaacetic acid). The stability of the common ammonia-carboxylation ligands with metal ions is shown in the appendix. EDTA is one of the most widely used titrants in analytical chemistry, and dozens of metal ions can be titrated with EDTA standard solution. Usually, the complexometric titration method is mainly referred to as EDTA titration.

6.1.3 Ethylenediaminetetra-acetic Acid (EDTA)

Ethylenediaminetetra-acetic acid (short for EDTA) designated H_4Y, the structure of EDTA is shown below:

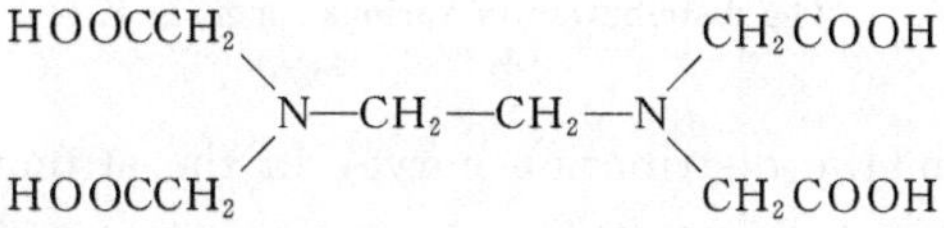

EDTA is a white crystalline powder with water-solubility lower at room temperature (22 ℃ when the solubility of 0.02 g/100 mL H_2O), insoluble in acid and organic solvents, soluble in alkali or corresponding salt formed in the ammonia. Due to EDTA's low solubility, it is unsuitable for titration.

EDTA disodium salt ($Na_2H_2Y \cdot 2H_2O$, also called EDTA, a relative molecular mass of 372.26) is a white crystalline powder that can absorb 0.3% moisture at room temperature, drying out at 80 ℃. During 100-140 ℃, $Na_2H_2Y \cdot 2H_2O$ might lose water and become anhydrous EDTA disodium salt (relative molecular mass is 336.24). EDTA disodium salt dissolves well in water (22 ℃ the solubility is 11.1 g/100 mL H_2O, concentration is about 0.3 mol $\cdot$ L^{-1} pH $\approx$ 4.4), so it is usually used as the titrant.

EDTA has a double dipole ion structure in an aqueous solution:

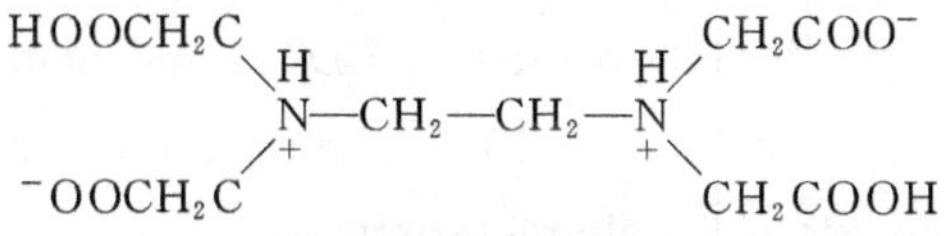

Therefore, when EDTA is dissolved in a solution with high acidity, the two carboxylate ions could be reaccepted as two H^+ so that it is equivalent to a six-element acid with a six-stage dissociation constant:

K_{a_1}	K_{a_2}	K_{a_3}	K_{a_4}	K_{a_5}	K_{a_6}
$10^{-0.9}$	$10^{-1.6}$	$10^{-2.0}$	$10^{-2.67}$	$10^{-6.16}$	$10^{-10.26}$

EDTA is always present in an aqueous solution of seven types, such as H_6Y^{2+}, H_5Y^+, H_4Y, H_3Y^-, H_2Y^{2-}, HY^{3-} and Y^{4-}, the relationship between the distribution coefficient δ and the pH of a solution is shown in Figure 6. 1.

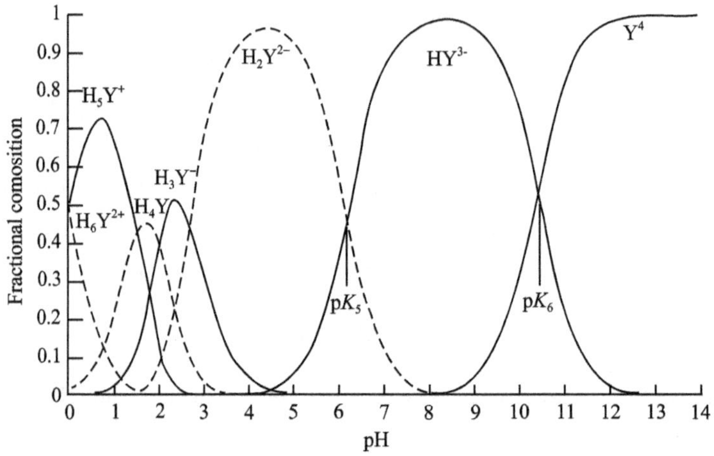

Figure 6. 1 The distribution of various forms in EDTA solution

As could be seen from the distribution curve, in the strong acid solution of pH<1, EDTA mainly exists in the form H_6Y^{2+}; when the pH is 2. 75-6. 24, the main type is H_2Y^{2-}; only in pH>10. 34 is the main type of presence Y^{4-}. It is important to note that only Y^{4-} (Y for Y^{4-}, usually omit the charges) could be directly coordinated with metal ions. The greater the δ_Y, the stronger complexation ability of EDTA, and the δ_Y is closely related to a solution's pH, so the solution's acidity is an important factor affecting the stability of EDTA-metal complexes and the acuity of the titration endpoint.

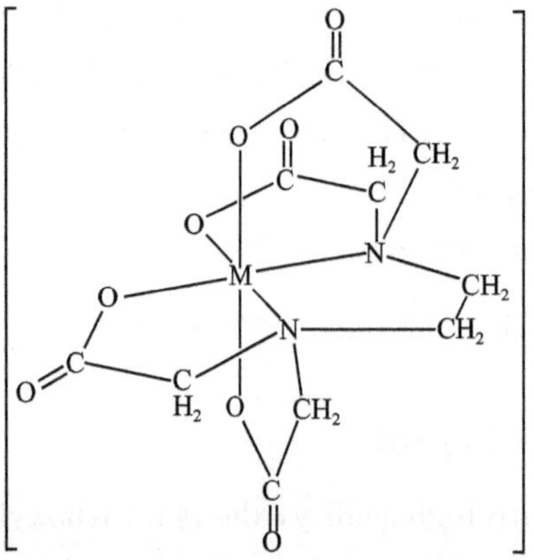

Figure 6. 2 EDTA-metal ion complex

6. 1. 4 EDTA-Metal Ion Complexes

In the EDTA molecule, there are six coordination atoms, which could form a five-ring chelate complex with one metal ion in the space position. Figure 6. 2 shows the three-dimensional structure of chelates formed by EDTA and metal ions (M).

EDTA-metal ion complex has the following characteristics:

(1) The EDTA has extensive coordination performance, which could form complexes with all metal ions. Thus, EDTA is widely used, but improving

its selectivity has become an important problem in complexometric titration.

(2) The coordination ratio of EDTA complexes is simple and, in most cases, 1 : 1. Individual metal ions such as Mo (Ⅴ) coordinated with EDTA to form complex $[(MoO_2)_2Y^{2-}]$.

(3) EDTA-metal ion complexes could form many five-atom ring structures with metal ions, which have high stability.

(4) EDTA-metal ion complexes are soluble in water, and the complexation reaction is rapid.

(5) Most EDTA-metal ion complexes are colorless, conducive to the endpoint's determination. However, the color of the chelates with nonferrous metal ions is darker. Therefore, when titrating these ions, their concentration must be controlled. Otherwise, using the indicator to determine the destination will be difficult. Such as:

CuY^{2-}	NiY^{2-}	CoY^{2-}	MnY^{2-}	CrY^-	FeY^-
dark blue	blue	purplish red	purplish-red	dark purple	yellow

6. 1. 5 Formation Constants of Complexes

1. The formation constant of complexes

For a simple complex, $M + L \rightleftharpoons ML$, the formation constant K_{ML} is shown below:

$$K_{ML} = \frac{[ML]}{[M] \times [L]} \qquad \textbf{(Equation 6. 1)}$$

K_{ML} can be expressed K_f. For the same type of complexes, the larger the value, the more stable the complexes are. The reciprocal of the formation constant is the in-stability constant of the complexes (K_d).

$$K_f = \frac{1}{K_d} \qquad \textbf{(Equation 6. 2)}$$

or:
$$\lg K_f = pK_d$$

The coordination ratio of most EDTA- metal ion complexes is 1 : 1 for $M+Y \rightleftharpoons MY$, so $K_f = \frac{[MY]}{[M]\times[Y]}$. The K_f of MY complexes can be seen in Table 6. 2. The K_f refers to the data in the case of no side reaction, so K_f does not reflect the stability of the complexes in the actual titration process.

2. Stepwise formation constant and cumulative constant

For the coordination ratio of 1 : n, the formation of ML_n is step by step, and the stepwise stability constant is K_{f_n}.

$$M+L \rightleftharpoons ML \qquad K_{f_1} = \frac{[ML]}{[M] \times [L]}$$

$$ML+L \rightleftharpoons ML_2 \qquad K_{f_2} = \frac{[ML_2]}{[ML] \times [L]}$$

and so on to:

Table 6.2 The formation constants for EDTA-metal ion complexes

Ion	lg K_f	Ion	lg K_f	Ion	lg K_f
Na^+	1.66	Ce^{4+}	15.98	Cu^{2+}	18.80
Li^+	2.79	Al^{3+}	16.30	Ga^{2+}	20.30
Ag^+	7.32	Co^{2+}	16.31	Ti^{3+}	21.30
Ba^{2+}	7.86	Pt^{2+}	16.31	Hg^{2+}	21.80
Mg^{2+}	8.69	Cd^{2+}	16.49	Sn^{2+}	22.10
Sr^{2+}	8.73	Zn^{2+}	16.50	Th^{4+}	23.20
Be^{2+}	9.20	Pb^{2+}	18.04	Cr^{3+}	23.40
Ca^{2+}	10.69	Y^{3+}	18.09	Fe^{3+}	25.10
Mn^{2+}	13.87	VO^+	18.10	U^{4+}	25.80
Fe^{2+}	14.33	Ni^{2+}	18.60	Bi^{3+}	27.94
La^{3+}	15.50	VO^{2+}	18.80	Co^{3+}	36.00

$$ML_{(n-1)} + L \rightleftharpoons ML_n \quad K_{f_n} = \frac{[ML_n]}{[ML_{n-1}]\times[L]} \qquad \textbf{(Equation 6.3)}$$

The cumulative constants, designated as β, are often used in dealing with association equilibria. The cumulative constant for the formation of ML_n is designated β_n and it is simply the product of the K_f values:

$$M + nL \rightleftharpoons ML_n \qquad \beta_n = K_{f_1}\times K_{f_2}\times\cdots\times K_{f_n} = \frac{[ML_n]}{[M]\times[L]^n} \qquad \textbf{(Equation 6.4)}$$

In a complex solution containing M, L, and various ML_n species, according to the cumulative constants β_n, the concentrations of all the ML_n can be calculated:

$$[ML] = K_{f_1}[M]\times[L] = \beta_1[M]\times[L]$$

$$[ML_2] = K_{f_2}[ML]\times[L] = K_{f_1}K_{f_2}[M]\times[L]^2 = \beta_2[M]\times[L]^2$$

And so on to:

$$[ML_n] = K_{f_n}[ML_{(n-1)}]\times[L] = K_{f_1}K_{f_2}\cdots K_{f_n}[M]\times[L]^n = \beta_n[M]\times[L]^n$$

(Equation 6.5)

Example 6.1

The formation constants of CaY^{2-} is $K(CaY^{2-}) = 10^{10.69}$, calculate the concentration of Ca^{2+} and pCa of a solution of 5.0×10^{-3} mol · L^{-1} at pH=12.

Solution:

The complex ion CaY^{2+} is sufficiently stable; the complexation reaction is virtually complete, so $[CaY^{2-}]-[Ca^{2+}]\approx[CaY^{2-}]=5.0\times10^{-3}$ mol · L^{-1}. At the complexation equilibrium, we have an equivalent dissociation ion of Ca^{2+} and Y. Thus, $[Ca^{2+}]=[Y]$. Let x represent $[Ca^{2+}]$:

$$[Ca^{2+}]+[Y^{4-}]\rightleftharpoons[CaY^{2-}]$$

$$x \qquad\qquad x \qquad\qquad 5.0\times10^{-3}-x\approx5.0\times10^{-3}$$

$$K(CaY^{2-})=\frac{[CaY^{2-}]}{[Ca^{2+}]\times[Y^{4-}]}=\frac{5.0\times10^{-3}}{x\times x}=10^{10.69}\Longrightarrow x=3.2\times10^{-7}$$

$\therefore c(Ca^{2+})=3.2\times10^{-7}$ mol · L^{-1}, $pCa=-\lg c(Ca^{2+})=6.5$

In the solution, the concentration of Ca^{2+} is 3.2×10^{-7} mol · L^{-1} and pCa is 6.5.

3. The distribution of all forms of the complexes in the solution

The influence of ligand concentration on the distribution of the complexes of all forms should be considered in the complexation equilibrium. If the concentration of metal ions is $c(M)$, the material balance of metal ions is below:

$$c(M)=[M]+[ML]+[ML_2]+\cdots+[ML_n]=[M](1+\sum_{i=1}^{n}\beta_i[L]^i) \quad \textbf{(Equation 6.6)}$$

The concentration of the complexes of all forms could be deduced by Equation 6.4. Therefore, the distribution coefficient of the complexes in all states is:

$$\delta(M)=\frac{[M]}{c(M)}=\frac{1}{1+\sum_{i=1}^{n}\beta_i[L]^i} \quad \textbf{(Equation 6.7)}$$

$$\delta(ML)=\frac{[ML]}{c(M)}=\frac{\beta_1[L]}{1+\sum_{i=1}^{n}\beta_i[L]^i} \quad \textbf{(Equation 6.8)}$$

$$\delta(ML_n)=\frac{[ML_n]}{c(M)}=\frac{\beta_n[L]^n}{1+\sum_{i=1}^{n}\beta_i[L]^i} \quad \textbf{(Equation 6.9)}$$

6.1.6 Side Reaction Coefficient and Conditional Formation Constant

In the process of titration, the reaction of EDTA (Y) with the measured metal ion (M) is described as the main reaction, while the other reactions in the solution are called side reactions, as follows:

Main reaction:

Side reactions:

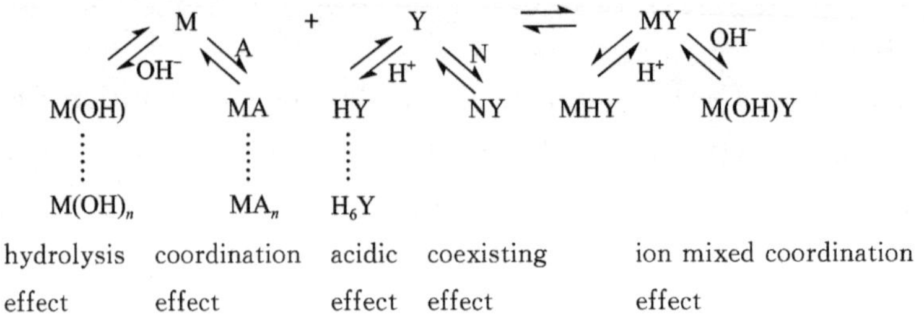

The side reaction of reactants (M and Y) goes against the main reaction, and the side reaction of products (MY) is good for the main reaction. But most side reactions of MY have poor stability and could be neglected. Therefore, the following discussion is on the side reactions of the reactants.

1. Side reaction coefficient

The concept of the side reaction coefficient is introduced to deal with the influence of various factors on complexation equilibrium. The side reaction coefficient measures the degree of influence on the side reaction to the main reaction, shown as α。

(1) Acidic effect and acidic effect coefficient ($\alpha_{Y(H)}$) Many ligands and chelating agents like EDTA have acid-base properties. The preceding exercise demonstrated that pH dramatically affects the active fraction of EDTA. We called this effect the "acidic effect", which is represented by the acidic effect coefficient ($\alpha_{Y(H)}$).

We already know that EDTA could exist in many forms through an acid-base reaction, with only Y^{4-} α form showing significant binding for most metal ions. $\alpha_{Y(H)}$ can be calculated from Equations 6.10.

$$\alpha_{Y(H)}=\frac{[Y']}{[Y]} \qquad \textbf{(Equation 6.10)}$$

Where $[Y']$ is for the concentration of all forms of EDTA not binding to metal ions, and $[Y]$ is for the concentration of free Y^{4-} of EDTA.

$$Y'=[Y]+[HY]+[H_2Y]+[H_3Y]+[H_4Y]+[H_5Y]+[H_6Y]$$

$$\alpha_{Y(H)}=\frac{[Y']}{[Y]}=\frac{[Y]+[HY]+[H_2Y]+[H_3Y]+[H_4Y]+[H_5Y]+[H_6Y]}{[Y]}$$

Put the six-stage dissociation constant of EDTA into the formula above:

$$\alpha_{Y(H)}=1+\frac{[H]}{K_{a_6}}+\frac{[H]^2}{K_{a_6}K_{a_5}}+\frac{[H]^3}{K_{a_6}K_{a_5}K_{a_4}}+\cdots+\frac{[H]^6}{K_{a_6}K_{a_5}\cdots K_{a_1}} \qquad \textbf{(Equation 6.11)}$$

$\alpha_{Y(H)}$ decreases with the increase of pH value. The smaller value of $\alpha_{Y(H)}$, the greater the effective concentration of $[Y]$, the less acidity will affect the complexes.

Example 6.2

Calculate the acid effect coefficient $\alpha_{Y(H)}$ of EDTA and the logarithmic value at pH=5, if the total concentration of EDTA is 0.02 mol·L^{-1}, please calculate the Y^{4-}.

Solution:

$$\alpha_{Y(H)}=1+\frac{10^{-5}}{10^{-10.26}}+\frac{10^{-10}}{10^{-10.26-6.16}}+\frac{10^{-15}}{10^{-10.26-6.16-2.67}}+\frac{10^{-20}}{10^{-10.26-6.16-2.67-2.0}}$$

$$+\frac{10^{-25}}{10^{-10.26-6.16-2.67-2.0-1.6}}+\frac{10^{-30}}{10^{-10.26-6.16-2.67-2.0-1.6-0.9}}\approx 10^{6.45}$$

$\therefore \lg \alpha_{Y(H)}=6.45$

$$[Y^{4-}]=\frac{[Y']}{\alpha_{Y(H)}}=\frac{0.02}{10^{6.45}}=7\times10^{-9}\ \text{mol}\cdot\text{L}^{-1}$$

$\alpha_{Y(H)}$ is the most used side reaction coefficient in EDTA titration. For the convenience of application, the numerical $\lg \alpha_{Y(H)}$ in different pH values are calculated using the method of Example 6.2. Table 6.3 lists the EDTA acid effect coefficient $\lg \alpha_{Y(H)}$ in different pH solutions.

Table 6.3 The EDTA acidic effect coefficient $\lg \alpha_{Y(H)}$ in different pH solutions

pH	$\lg \alpha_{Y(H)}$	pH	$\lg \alpha_{Y(H)}$	pH	$\lg \alpha_{Y(H)}$	pH	$\lg \alpha_{Y(H)}$
0.0	23.64	3.8	8.85	7.0	3.32	10.0	0.45
0.4	21.32	4.0	8.44	7.4	2.88	10.4	0.24
0.8	19.08	4.4	7.64	7.8	2.47	10.8	0.11
1.0	18.01	4.8	6.84	8.0	2.27	11.0	0.07
1.4	16.02	5.0	6.45	8.4	1.87	11.4	0.03
1.8	14.27	5.4	5.69	8.8	1.48	12.0	0.01
2.0	13.51	5.8	4.98	9.0	1.28	13.0	0.000 8
2.4	12.19	6.0	4.65	9.5	0.83	13.9	0.000 1

(2) Coordination effect and coordination effect coefficient ($\alpha_{M(L)}$) In EDTA titration, other ligands binding to metal ions, called the coordination effect, could reduce the main reaction. The side effect coefficient, which is caused by the other ligands (L), is called the coordination effect coefficient $\alpha_{M(L)}$, defined the ratio of the total concentration of metal ions [M'] to the free metal ion concentration [M]. The value of $\alpha_{M(L)}$ could be deduced by Equations 6.4 and 6.5:

$$\alpha_{M(L)}=\frac{[M']}{[M]}=\frac{[M]+[ML_1]+[ML_2]+\cdots+[ML_n]}{[M]}$$

$$=1+\beta_1[L]+\beta_2[L]^2+\cdots+\beta_n[L]^n \qquad \textbf{(Equation 6.12)}$$

The larger value of $\alpha_{M(L)}$, the more severe side reaction effect. So, we usually use a buffer to prevent metal ion hydrolysis by other ligands "L", or a masking agent to eliminate interference.

(3) Side reaction of complexes MY This side effect occurs at high acidity or low

acidity. When the acidity is high, the acid compound (MHY) is formed, and its side reaction coefficient is expressed $\alpha_{MY(H)}$; when the acidity is low, the alkali-type complexes (MOHY) are produced, and the side reaction coefficients are expressed $\alpha_{MY(OH)}$. Acid and alkali complexes are generally unstable and often negligible in the calculation.

2. Conditional formation constant

We have discussed those side reaction effects of the main reaction, so the absolute formation constant cannot describe the stability of the complexes at all. Considering the influence of side reaction effects, the **conditional formation constant** K'_{MY} should be introduced. The relationship between K'_{MY} and $\alpha_{Y(H)}$ and $\alpha_{M(L)}$ is below (ignoring the side reaction effects of MY):

$$K'_{MY}=\frac{[MY']}{[M']\times[Y']} \qquad \textbf{(Equation 6.13)}$$

$$\because K_{MY}=\frac{[MY]}{[M]\times[Y]},\ \alpha_{Y(H)}=\frac{[Y']}{[Y]},\ \alpha_{M(L)}=\frac{[M']}{[M]}$$

$$\therefore K'_{MY}=\frac{[MY]}{[M]\alpha_{M(L)}\times[Y]\alpha_{Y(H)}}\Rightarrow K'_{MY}=\frac{K_{MY}}{\alpha_{M(L)}\times\alpha_{Y(H)}}$$

$$\therefore K'_{MY}=\frac{[MY]}{[M]\alpha_{M(L)}\times[Y]\alpha_{Y(H)}}\Rightarrow K'_{MY}=\frac{K_{MY}}{\alpha_{M(L)}\times\alpha_{Y(H)}} \qquad \textbf{(Equation 6.14)}$$

Take the logarithm of Equation 6.14:

$$\lg K'_{MY}=\lg K_{MY}-\lg \alpha_{M(L)}-\lg \alpha_{Y(H)} \qquad \textbf{(Equation 6.15)}$$

Conditional formation constant, K'_{MY}, is the actual stability constant of EDTA-metal ion complexes, which could be used to calculate complexometric titration, complexation equilibrium, and metal ion concentration.

Example 6.3

Calculate the $\lg K'_{ZnY}$ at pH=2.00 and pH=5.00, respectively.

Solution:

For Table 6.2, $\lg K_{ZnY}=16.50$; for Table 6.3, pH=2.0, $\lg \alpha_{Y(H)}=13.51$; pH=5.0, $\lg \alpha_{Y(H)}=6.45$.

If only consider the acidic effect, $\lg K'_{ZnY}=\lg K_{ZnY}-\lg \alpha_{Y(H)}$

pH=2.0, $\lg K'_{ZnY}=\lg K_{ZnY}-\lg \alpha_{Y(H)}=16.50-13.51=2.99$

pH=5.0, $\lg K'_{ZnY}=\lg K_{ZnY}-\lg \alpha_{Y(H)}=16.50-6.45=10.05$

It can be seen from the above example although $\lg K_{ZnY}=16.50$, when pH=2.0, $\lg K'_{ZnY}$ is only 2.99, and ZnY^{2-} is unstable, the Zn^{2+} cannot be titrated accurately in this condition; At pH=5.0, $\lg K'_{ZnY}$ is 10.05, ZnY^{2-} is stabilized, and the titration can be carried out. It is very important to control the acidity of a solution.

6.2 Metal Ion Indicators

6.2.1 Principle of Metal Ion Indicator

The most common method to detect the endpoint in EDTA titration is metal ion indicators. Metal ion indicator is an organic dye, also a coordination agent, which could react with metal ions to produce a complex that is a significantly different color from the EDTA-metal ion complex to indicate the endpoint.

Metal ion indicators are compounds whose color changes when they bind to a metal ion. Useful indicators must bind to the metal ion less strongly than EDTA does, $K'_{MY}/K'_{MIn} \geqslant 10^2$ at least. A typical titration is illustrated by an EDTA (Y) reaction with Mg^{2+}, using Eriochrome Black T as the indicator.

$$\underset{\text{red}}{MgIn} + \underset{\text{colorless}}{Y} \rightleftharpoons \underset{\text{colorless}}{MgY} + \underset{\text{blue}}{In}$$

At the start of the experiment, a small number of Eriochrome black T (symbol for In) are added to the colorless solution of Mg^{2+}, forming red complexes (MgIn). When EDTA (Y) is added, it reacts first with free Mg^{2+} ions. As free Mg^{2+} is used up, the last EDTA added before the equivalence point displaces Eriochrome Black T from the red MgIn complexes. The change from the red MgIn to blue unbound signals the endpoint of the titration.

6.2.2 Theoretical Discoloration Point of Metal Ion Indicator

If the metal ion indicator is formed with metal ions by 1 : 1, the reaction is:

$$M + In \rightleftharpoons MIn$$

Consider the acidic effect of the indicator:

$$K'_{MIn} = \frac{[MIn]}{[M] \times [In']} \Longrightarrow \lg K'_{MIn} = pM + \lg \frac{[MIn]}{[In']}$$

Like the acid-base indicator, when $[MIn] = [In']$, the solution presents a mixed color of MIn and In. pM_t is the theoretical discoloration point of the metal indicator.

$$pM_t = \lg K'_{MIn} = \lg K_{MIn} - \lg \alpha_{In(H)} \qquad \textbf{(Equation 6.16)}$$

The metal ion indicators are weak acid-base that have an acidic effect. Equation 6.16 indicates that the $\lg K'_{MIn}$ of the indicator varies with the pH change, and it cannot have a definite color like the acid-base indicator. Therefore, the system's acidity should be considered when selecting an indicator so that the discoloration point pM_t is as close as possible to the stoichiometric point. In practical work, we mostly adopt the experiment to choose the appropriate indicator.

6.2.3 Common Metal Ion Indicator

Most metal ion indicators are acid-base compounds, with pK_a listed in Table 6.4 because the color of free indicators depends on the pH. Most indicators could be used only in certain pH ranges. For example:

Eriochrome Black T, short for EBT, there is a balance in the solution:

$$\begin{array}{ccc} & pK_{a_2}=6.3 & & pK_{a_3}=11.6 & \\ H_2In & \rightleftharpoons & HIn^{2-} & \rightleftharpoons & In^{3-} \\ \text{red} & & \text{blue} & & \text{orange} \end{array}$$

When $pH<6.3$, EBT is red in an aqueous solution, while $pH>11.6$, EBT is orange. The complex of EBT with bivalent metal ions is red-violet, only in the pH range of 7-11, the indicator's color is obvious. The experiments show that the optimum acidity is from 9 to 10.5 pH.

The solid Eriochrome Black T is stable, but its aqueous solution can only be kept for a few days due to the polymerization reaction. After polymerization, Eriochrome Black T cannot be coordinated with metal ions. Therefore, Eriochrome Black T is usually prepared to mix with solid NaCl by the weight ratio of 1 : 100. EBT is a common indicator for titrating Mg^{2+}, Zn^{2+}, and Pb^{2+} in a weak alkaline solution.

The xylenol orange changes from yellow to red when it binds to a metal ion at pH=5.5, which is an easy phenomenon to observe. At pH=7.5, the change is from violet to red.

Table 6.4 Common metal ion indicator

Name	pK_a	Color of free indicator	Color of metal ion complex	Titrated metal ions	Metal ions block the indicator
Acid chrome blue K	$pK_{a_1}=6.7$ $pK_{a_2}=10.2$ $pK_{a_3}=14.6$	H_2In^- red HIn^{2-} blue In^{3-} orange	Red	Mg^{2+} (pH 10) Ca^{2+} (pH 12)	
Calmagite	$pK_{a_2}=3.8$ $pK_{a_3}=9.4$ $pK_{a_4}=13\text{-}14$	H_2In^- red HIn^{2-} blue In^{3-} orange	Wine red	Ca^{2+} (pH 12-13)	Co^2+, Ni^{2+}, Cu^{2+}, Fe^{3+}, Al^{3+}, Ti^{4+}
Eriochrome Black T	$pK_{a_1}=3.9$ $pK_{a_2}=6.3$ $pK_{a_3}=11.6$	H_2In^- red HIn^{2-} blue In^{3-} orange	Wine red	Ca^{2+} (pH 10) Mg^{2+} (pH 10) Pb^{2+} (pH 10) Zn^{2+} (pH 6.8-10)	Co^{2+}, Ni^{2+}, Cu^{2+}, Fe^{3+} Al^{3+}, Ti^{4+}
Murexide	$pK_{a_1}=1.6$ $pK_{a_2}=9.2$ $pK_{a_3}=10.9$	H_4In^- red-violet H_3In^{2-} violet H_2In^{3-} blue	Yellow (with Ni^{2+}, Cu^{2+}) Red with Ca^{2+}	Ca^{2+} (pH>10) Cu^{2+} (pH 7-8) Ni^{2+} (pH 8.5-11.5)	

Table 6.4 continued

Name	pK_a	Color of free indicator	Color of metal ion complex	Titrated metal ions	Metal ions block the indicator
o-PAN	$pK_{a_1}=1.9$ $pK_{a_2}=12.2$	HIn^- yellow In^{2-} yellow	Red Pink	Cu^{2+} (pH 6) Zn^{2+} (pH 5-7)	
Ulfosalicylic acid	$pK_{a_1}=2.6$ $pK_{a_2}=11.7$	HIn^- yellow	Purple	Fe^{3+} (pH 1.5-3)	
Xylenol orange	$pK_{a_1}=1.6$ $pK_{a_2}=2.33$ $pK_{a_3}=2.85$ $pK_{a_4}=6.70$ $pK_{a_5}=10.47$ $pK_{a_6}=12.33$	H_5In^- yellow H_4In^{2-} yellow H_3In^{3-} yellow H_2In^{4-} violet HIn^{5-} violet In^{6-} violet	Red	Zn^{2+}, Pb^{2+}, Hf^{3+}, Al^{3+}, Ni^{2+}, Co^{2+}, Th^{3+}, Sc^{3+}, Bi^{2+}, Cu^{2+}, Ga^{2+} Hg^{2+}, Cd^{2+}, In^{2+} (pH 0-6) Fe^{3+} (pH 2-3)	

6.2.4 Announcements of Using Metal Ion Indicator

1. Blocking of indicator

For an indicator to be useful in titrating metal ions by EDTA, the indicator must give up its metal ion to EDTA. If a metal ion does not freely dissociate from an indicator, the metal ion is said to block the indicator due to lg K_{MIn}>lg K_{MY}. For example, Eriochrome Black T is blocked by Cr^{3+}, Fe^{3+}, Cu^{2+}, Ni^{2+}, Co^{2+}, and Al^{3+}. We can add a masking agent, such as Triethanolamine, so that the interfering ions generate more stable complexes and no longer interact with the indicator. For example, Cu^{2+}, Co^{2+}, and Ni^{2+} could use KCN for masking, While Fe^{3+} is required to reduce to Fe^{2+} by ascorbic acid and then KCN masking.

If there are no useful masking agents or indicators for direct titration, it can be used for back titration. For example, an excess standard EDTA solution is added to Cu^{2+}, and then an indicator is added to the excess EDTA and back-titrated with Zn^{2+}. If the number of interfering ions is too many, the ions must be removed.

2. Ossification of indicator

The low solubility of indicators or metal-indicator complex (MIn) in the water could lead to the slow exchange process of titration ligands (L) with metal complexes (MIn), which might then extend the endpoint. This phenomenon is called **ossification.** The solution is to add organic solvent or heat to increase its solubility. For example, when using PAN as an indicator, alcohol or titrate are often added under heating.

3. Oxidative metamorphism of the indicator

The solution of azo indicators (compounds with —N=N— bonds) deteriorates rapidly and should probably be prepared each week. Murexide solution should be prepared daily.

Most metal ion indicators are azo indicators (compounds with —N=N— bonds), which deteriorate rapidly by sunlight, oxidizer, and air and are unstable in aqueous solutions. Murexide solution should be prepared daily. The solid mixture is more stable than the solution and has a longer storage time. For example, Eriochrome Black T and Calmagite commonly use solid NaCl or KCl as diluents to prepare.

6.3 Complexometric Titration Curve

In a complexometric titration, with the addition of ligands at certain pH conditions, the metal ions continuously react with ligands to generate complexes, and their concentration decreases. The metal ion concentration (pM) suddenly changes when titration reaches the stoichiometric point. We take the volume of the ligand as the horizontal coordinate and metal ion concentration (pM) as the vertical coordinate to obtain the complexometric titration curve. The relationship between the number of ligands and the concentration of the metal ions is reflected in the titration curve.

6.3.1 Curve Drawing

Construct the titration curve of pCa vs. volume of EDTA for 20.0 mL of 0.010 00 mol · L^{-1} Ca^{2+} being titrated with 0.010 00 mol · L^{-1} EDTA in a solution buffered to a constant pH of 12.0.

Calculating the conditional formation constant:

The conditional formation constant of the Ca-EDTA complex at pH 12.0 is obtained from the formation constants of the complex (Table 6.2) and the $\alpha_{Y(H)}$ for EDTA at pH 12.0. If we substitute them into Equation 6.15, we could get:

$$\lg K'_{CaY} = \lg K_{CaY} - \lg \alpha_{Y(H)} = 10.7 - 0.01 = 10.69$$

(1) Pre-titration $[Ca^{2+}] = 0.010\ 00\ mol \cdot L^{-1}$, pCa=2.00。

(2) Before the stoichiometric point The residual metal ion Ca^{2+} and complexes CaY are in the solution. Due to the larger $\lg K'_{CaY}$, the remaining Ca^{2+} has a certain inhibitory effect on the dissociation of CaY, which can ignore and calculate the pCa according to the remaining metal ion concentration $[Ca^{2+}]$.

$$\text{When } V(\text{EDTA}) = 18.00\ \text{mL}: [Ca^{2+}] = \frac{2.00 \times 0.010\ 00}{20.00 + 18.00} = 5.26 \times 10^{-4}\ (mol \cdot L^{-1})$$

$$pCa = -\lg[Ca^{2+}] = 3.28$$

$$\text{When } V(\text{EDTA}) = 19.98\ \text{mL}: [Ca^{2+}] = \frac{0.010\ 00 \times 0.02}{20.00 + 19.98} = 5 \times 10^{-6}\ (mol \cdot L^{-1})$$

$$pCa = -\lg[Ca^{2+}] = 5.3$$

(3) Stoichiometric point When $V(\text{EDTA}) = 20.00$ mL, Ca^{2+} and EDTA completely generate the complexes CaY.

$$[CaY]=0.010\ 00\times\frac{20.00}{20.00+20.00}=5\times10^{-3}(mol\cdot L^{-1})$$

Meanwhile, $[Ca^{2+}]=[Y^{4-}]$:

$$K'_{MY}=\frac{[CaY]}{[Ca^{2+}]\times[Y]}=\frac{[CaY]}{[Ca^{2+}]^2}\Rightarrow\frac{5\times10^{-3}}{[Ca^{2+}]^2}=10^{10.69}\Rightarrow[Ca^{2+}]=3.2\times10^{-7}$$

$$pCa=-\lg[Ca^{2+}]=6.5$$

(4) After stoichiometric point When $V(EDTA)=20.02$ mL, excess EDTA standard solution is 0.02 mL.

$$[Y]=\frac{0.010\ 00\times0.02}{20.00+20.02}=5\times10^{-6}(mol\cdot L^{-1})$$

$$K'_{MY}=\frac{[CaY]}{[Ca^{2+}]\times[Y]}=\frac{5\times10^{-3}}{[Ca^{2+}]\times5\times10^{-6}}=10^{10.69}\Rightarrow[Ca^{2+}]=10^{-7.69}$$

$$pCa=-\lg[Ca^{2+}]=7.69$$

The calculation results of the whole titration process are shown in Table 6.5. The stoichiometric point of pCa is 6.5, and the titration jump of pCa is 5.3-7.7. The titration jump is larger and could be titrated accurately.

Table 6.5 The pCa at pH=12, 0.010 00 mol · L^{-1} EDTA titrated 20.00 mL 0.010 00 mol · L^{-1} Ca^{2+}

V(EDTA)/mL	EDTA/%	Titrated Ca^{2+}/%	Excess EDTA/%	pCa	
0	0			2.0	
18.00	90.0	90.0		3.3	
19.80	99.0	99.0		4.3	
19.98	99.9	99.9		5.3	Titration jump
20.00	100.0	100.0		6.5	Titration jump
20.02	100.1		0.1	7.7	Titration jump
20.20	101.0		1.0	8.7	
40.00	200.0		100	10.7	

According to the data in Table 6.5, the volume of EDTA as the horizontal coordinate, the pCa as the vertical coordinate, and the titration curve of Figure 6.3 is obtained.

6.3.2 Factors Affecting the Range of Titration Jump

The larger the titration jump, the easier it is to indicate the titration endpoint accurately. The above example results show that the conditional formation constants and the concentration of titration metal ions are the main factors influencing the jump range.

1. The conditional formation constants ($\lg K'_{MY}$)

Figure 6.4 shows the titration curve of different $\lg K'_{MY}$. We could see the larger $\lg K'_{MY}$, and larger the titration jump (ΔpM). Factors determining the $\lg K'_{MY}$ are the absolute formation constant $\lg K'_{MY}$, the solution acidity, coordination of masking agent, and other

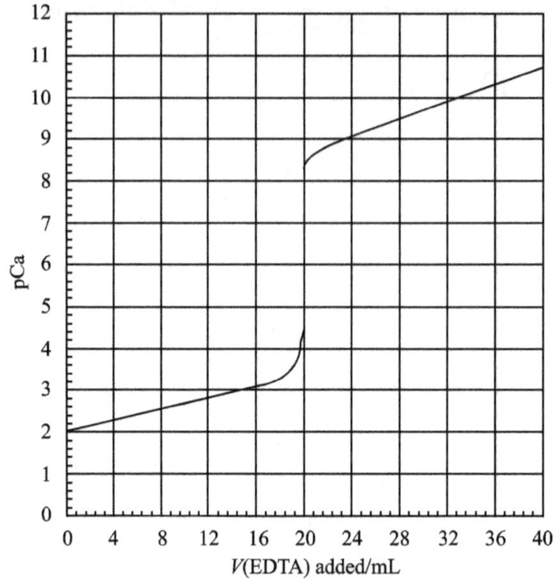

Figure 6.3 At pH=12.0, the titration curve of 0.010 00 mol · L^{-1} EDTA titrated 0.010 00 mol · L^{-1} Ca^{2+}

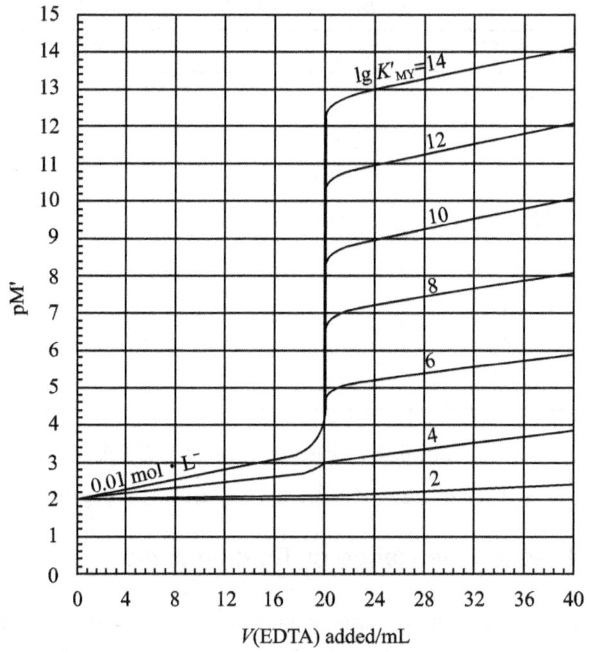

Figure 6.4 Titration curve for 0.010 00 mol · L^{-1} solution of various lg K'_{MY} at pH 6.0

auxiliary complexing agents.

(1) Acidity, due to Equation 6.15, the higher the acidity, the larger the value of lg $\alpha_{Y(H)}$, the lg K'_{MY} becomes smaller, the titration jump gets smaller, as shown in Figure 6.5.

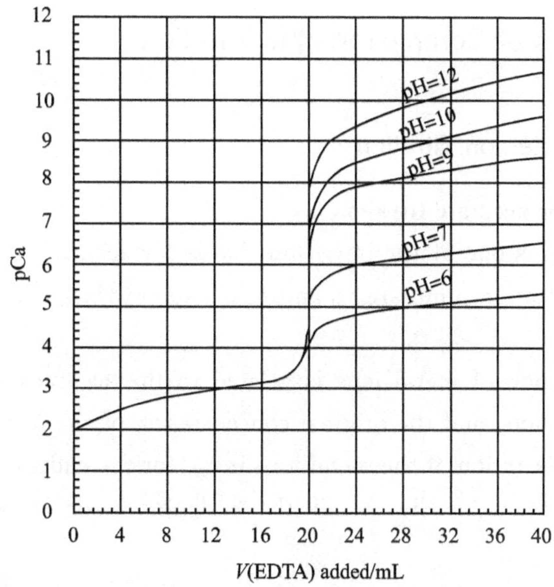

Figure 6. 5 Titration curve for 0. 010 00 mol · L^{-1} M solution of various pH

(2) The effects of other coordination agents on the addition of masking agents and buffer solutions will increase the lg $\alpha_{M(L)}$ and reduce the lg K'_{MY} titration jump to get smaller.

2. Concentration of metal ions

Figure 6. 6 is the titration curve of different metal ion concentrations. It can be seen that the titration jump gets wider when the concentration of metal ions is larger and the starting point of the titration curve is lower. At the same time, we know that the change only affects the front side of the titration curve.

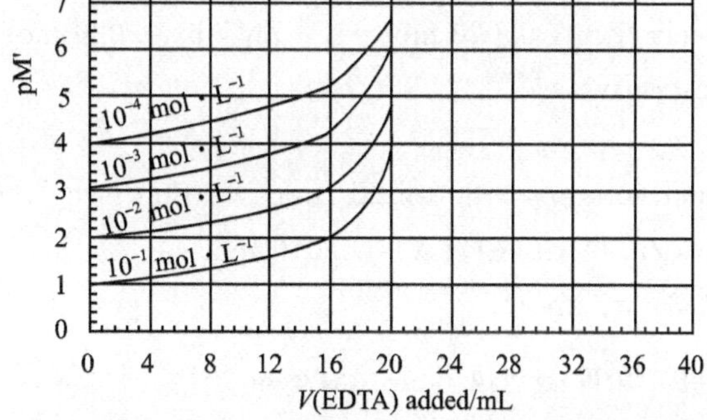

Figure 6. 6 The titration curve of different concentrations of metal ions

6.4 Conditions of Complexometric Titration

6.4.1 Single Metal Ion Titration

1. Discriminant for accurate titration

The titration jump is one of the important bases for accurate titration, while the main factors affecting the titration jump are c_M and K'_{MY}, so how large of c_M and K'_{MY} could make the titration accurate.

The accurate titration of metal ions is related to the accuracy of allowable error, the detection endpoint method, and the original concentration of the metal ions.

The original concentration of the metal ions is c_M (for the endpoint volume), and titrated by the same concentration of EDTA. If the endpoint error is less than 0.1% at the stoichiometric point:

(1) Almost all of the metal ions complexed with EDTA and formed MY, so $[MY]=c_M$.

(2) c_M is equal to $c(EDTA)$ at the stoichiometric point.

(3) Due to TE $\leqslant$ 0.1%, $c_{M,residue} \leqslant 0.1\% c_M$, $c(EDTA, residue) \leqslant 0.1\% c(EDTA)$:

$$\because K'_{MY}=\frac{[MY]}{c_{M,residue}\times c(EDTA, residue)}$$

$$\therefore K'_{MY}\geqslant\frac{c_M}{c_M\times 0.1\%\times c(EDTA)\times 0.1\%}\Rightarrow K'_{MY}\geqslant\frac{1}{c_M\times 0.1\%\times 0.1\%}$$

$$K'_{MY}\times c_M\geqslant\frac{1}{0.1\%\times 0.1\%}\Rightarrow \lg(K'_{MY}\times c_M)\geqslant 6$$

Discriminant of single metal ion titration is:

$$\lg(c_M\times K'_{MY})\geqslant 6 \qquad \textbf{(Equation 6.17)}$$

Example 6.4

Could we accurately titrate 0.020 mol·L^{-1} Zn^{2+} by 0.020 mol·L^{-1} EDTA at pH=2.0 and pH=5.0, respectively?

Solution:

Since we add equal volumes, the initial concentration is halved by dilution. At the stoichiometric point, $c(Zn^{2+})=c(EDTA)=0.010$ mol·L^{-1}.

From Example 6-3, we know:

$$pH=2.0,\ \lg K'_{ZnY}=\lg K'_{ZnY}-\lg \alpha_{Y(H)}=16.50-13.51=2.99$$

$$\lg(K'_{ZnY}\times c_M)=\lg K'_{ZnY}+\lg c_M=2.99-2=0.99<6$$

$$pH=5.0,\ \lg K'_{ZnY}=\lg K_{ZnY}-\lg \alpha_{Y(H)}=16.50-6.45=10.05$$

$$\lg(K'_{ZnY}\times c_M)=\lg K'_{ZnY}+\lg c_M=10.05-2=8.05>6$$

Therefore, when pH = 2.0, the Zn^{2+} is not accurately titrated, but at pH = 5.0 could be.

We can see from this example that, with EDTA titration of metal ions, it is necessary to choose the appropriate pH for titration because acidity is an important factor in accurately titrating metal ions.

2. Range of pH for a kind of metal ion titration

(1) The highest acidity (or the minimum pH) According to Equation 6.17, if we only consider the acidic effect in the titration reaction, when the concentration of the metal ion is 0.01 mol · L^{-1}, therefore $\lg K'_{MY} \geqslant 8$:

$$\lg K'_{MY} = \lg K_{MY} - \lg \alpha_{Y(H)} \geqslant 8$$

$$\lg \alpha_{Y(H)} \leqslant \lg K_{MY} - 8 \qquad \textbf{(Equation 6.18)}$$

For example, the concentration of Zn^{2+} is 0.01 mol · L^{-1}, $\lg K_{ZnY} = 16.50$, according to the Equation 6.18:

$$\lg \alpha_{Y(H)} \leqslant \lg K_{ZnY} - 8 \Rightarrow \lg \alpha_{Y(H)} \leqslant 16.50 - 8 \Rightarrow \lg \alpha_{Y(H)} \leqslant 8.50 \Rightarrow pH \geqslant 4.0$$

From Table 6.3, we could find that $\lg \alpha_{Y(H)} = 8.44$ at pH = 4.0, so the minimum pH of titration of Zn^{2+} is 4.0. Figure 6.7 is the effective acidic curve with $c_M = 0.010\ 00$ mol · L^{-1}, which shows the minimum pH at which EDTA could titrate different metal ions. The points on the curve represent the pH at which the conditional formation constant $\lg K'_{MY}$ for each metal, the pH is 8, which is arbitrarily chosen as the minimum pH required for a sharp end point. Note that the smaller the $\lg K'_{MY}$, the more alkaline the solution must be to obtain $\lg K'_{MY}$ of 8.

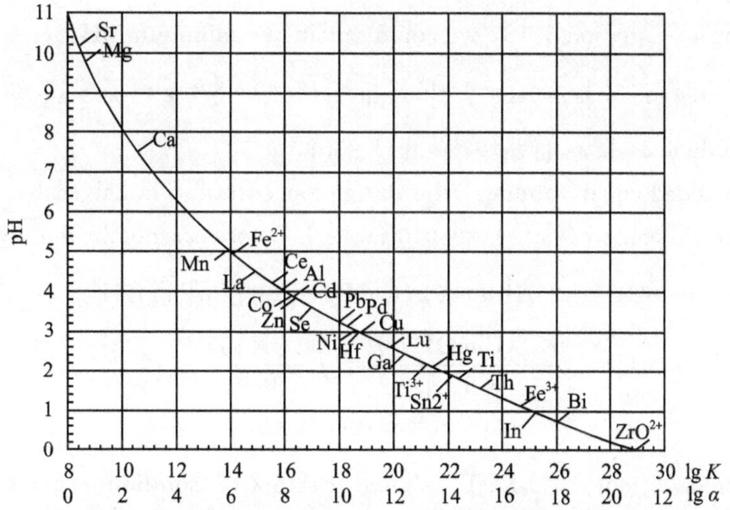

Figure 6.7 Minimum pH for effective acidic titration of various metal ions with EDTA (c_M = 0.010 00 mol · L^{-1})

(2) The lowest acidity (or the highest pH) Most metal ions could react at the high pH range, but not all metals could be titrated directly due to hydroxides $M(OH)n$ precipitation. For example, we have calculated the minimum pH of titration of Zn^{2+}, but to prevent the precipitation of $Zn(OH)_2$, it must be satisfied in the following formula:

$$Zn(OH)_2 \rightleftharpoons Zn^{2+} + 2OH^- \Rightarrow K_{sp}(Zn(OH)_2) = [Zn^{2+}][OH^-]^2$$

$$[OH] = \sqrt{\frac{K_{sp}(Zn(OH)_2)}{[Zn^{2+}]}} = \sqrt{\frac{2.09\times10^{-16}}{0.01}} = 1.4\times10^{-7}$$

$$pOH = 6.85 \Rightarrow pH = 7.15$$

Therefore, the 0.01 mol · L^{-1} Zn^{2+} solution should be titrated by EDTA in the range of pH 4.0-7.15. The higher the pH, the larger the titration jump would be. The pH will be higher if the auxiliary complexing agent is added to the metal ion solution (such as ammonia, tartaric acid, etc.). For example, the Zn^{2+} can be titrated in an ammonia buffer solution at pH=10.

We have discussed the range of suitable acidity for the titration reaction but also need to use indicators to indicate the titration endpoint. The metal indicators could only be used in a certain pH range. Due to the acidic effect, the indicator discoloration point is not fixed and is changed by the pH. Therefore, the choice of an indicator system of pH must be considered. In theory, the best acidity is when the stoichiometric point is closest to the indicator discoloration point, but we still need to determine the optimal acidity of the experiment.

Example 6.5

Calculate the pH range of 0.020 mol · L^{-1} EDTA titrated 20 mL 0.020 mol · L^{-1} Cu^{2+} solution. [$\lg K_{CuY} = 18.80$, $K_{sp}(Cu(OH)_2) = 4.8\times10^{-20}$]

Solution:

According to Equation 6.18, we could obtain the minimum pH:

$$\lg \alpha_{Y(H)} \leqslant \lg K_{ZnY} - 8 \Rightarrow \lg \alpha_{Y(H)} \leqslant 18.80 - 8 \Rightarrow \lg \alpha_{Y(H)} \leqslant 10.80$$

From Table 6.3, $\lg \alpha_{Y(H)} = 10.80$, pH=2.9.

Since we added equal volumes, the initial concentration is halved by dilution, so at the stoichiometric point, $c(Cu^{2+}) = 0.010$ mol · L^{-1}, we can obtain the highest pH:

$$Cu(OH)_2 \rightleftharpoons Cu^{2+} + 2OH^- \Rightarrow K_{sp}(Cu(OH)_2) = [Cu^{2+}][OH^-]^2 = 4.8\times10^{-20}$$

$$[OH] = \sqrt{\frac{K_{sp}(Cu(OH)_2)}{[Cu^{2+}]}} = \sqrt{\frac{4.8\times10^{-20}}{0.01}} = 2.2\times10^{-9}$$

$$pOH = 8.7 \Rightarrow pH = 5.3$$

Therefore 0.02 mol · L^{-1} EDTA should titrate Zn^{2+} solution in pH 2.9-5.3.

3. The role of buffer in coordination titration

In the process of coordination titration, it's going to release H^+, shown below:

$$M^{n+} + H_2Y^{2-} \rightleftharpoons MY^{(4-n)-} + 2H^+$$

The solution's acidity increases, affecting the main reaction and decreasing the indicator's sensitivity. Therefore, the acidity of the solution should be controlled by adding a buffer.

In weak acid conditions (pH 5-6), the acetic acid or hexamethylenetetramine solution buffer is often used. Ammonia buffer is usually used in weak alkaline conditions (pH 8-10). When choosing a buffer, not only the range of pH but also whether the buffer would cause the side reaction of the metal ions should be considered. For example, Pb^{2+} is titrated by EDTA at pH=5. We do not use acetic acid buffer because the Ac^- could be complexed with Pb^{2+}, and lower the $\lg K'_{PbY}$. In addition, the selected buffer solution must have sufficient buffer capacity to control the pH of the solution.

6.4.2 Selective Titration of Mixed Ions

EDTA could form stable complexes with most metal ions. However, it is more likely to be interfered with by other ions in EDTA titration. In the coexistence of various metal ions (N), reducing the interference of other ions to the main metal ion(M) and improving the selectivity are important problems to be solved in coordination titration.

If the coexistence ion N is not disturbed, then:

$$\frac{c_M \times K'_{MY}}{c_N \times K'_{NY}} = \frac{10^6}{c_N \times K'_{NY}} \geqslant 10^5$$

$$c_N \times K'_{NY} = 10^1 \Rightarrow \lg c_N \times K'_{NY} \leqslant 1 \qquad \textbf{(Equation 6.19)}$$

However, most interference ions are not satisfied, so the interference ions must be processed. Therefore, the following methods are generally adopted.

1. Control the acidity of the solution

Adjusting the pH of the solution and interference could change the $\lg K'_{MY}$ and $\lg K'_{NY}$. Using the acid effect curve can easily solve these problems. For example, when the sample of $ZnSO_4$ is titrated, it could be used not only for xylenol orange at pH=5-6, but also for Eriochrome Black T at pH=10 as an indicator. If $MgSO_4$ is present in the sample, the Zn^{2+} must be titrated only at pH=5-6, because at pH=10, Mg^{2+} would disturb the titration.

2. Masking agent

A **masking agent** is a reagent that protects some analyte components from reaction with EDTA. When several metal ions coexist, a masking agent could form a stable complex with interference ions that could be better masked. For example, in determining the content of Ca^{2+} and Mg^{2+} in water, triethanolamine can prevent the interference of Fe^{3+} and Al^{3+}.

The coordination masking method is applied to the coordination titration and is widely used to improve the selectivity of other titration reactions. Common masking agents include NH_4F, KCN, triethanolamine, and tartaric acid.

3. Oxidation-reduction method

Change the oxidation number of interference ions to eliminate the interference. For example, when we measure Bi^{3+}, if the Fe^{3+} could interfere with the main reaction, add vitamin C or hydroxylamine hydrochloride to make Fe^{3+} to form Fe^{2+} and eliminate the interference of Fe^{3+}.

6.5 EDTA Titration Methods

This section will introduce several titration methods because EDTA titration can analyze many elements.

6.5.1 Direct Titration

In **direct titrations**, the analyte is titrated with a standard EDTA solution. The analysis is buffered to a pH, where the $\lg K'_{MY}$ of the metal-EDTA complex is large, and the color of the free indicator is distinctly different from the metal-indicator complex.

Water hardness is the total concentration of alkaline mineral ions in water equal to $[Ca^{2+}]$ and $[Mg^{2+}]$ because the concentration of Mg^{2+} and Ca^{2+} is usually more than other alkaline mineral ions. Water hardness is expressed as the number of milligrams of $CaCO_3$ per liter. For example, $[Ca^{2+}]+[Mg^{2+}]=0.001\ mol\cdot L^{-1}$, we would say the total hardness is 100 mg $CaCO_3$ per liter. If hardness is above 270 $mg\cdot L^{-1}$, the water is said to be "hard water".

To measure the total hardness of Ca^{2+} and Mg^{2+}, the water is treated with a pH=10 ammonia buffer, then Eriochrome Black T as an indicator titrated by EDTA solution, usually using cyanide to mask Fe^{2+}, Cu^{+}, and several other minor metal ions. The concentration of Ca^{2+} could be determined separately if the titration is carried out at pH = 12 by adding NaOH, while Mg^{2+} is precipitated and inaccessible to the EDTA. The indicator is Calmagite. The right choice of indicators can reduce interference by many metal ions.

6.5.2 Back Titration

A measured excess of EDTA solution is added to the analyte in back titrations. After the reaction, the excess EDTA is titrated with another standard solution of a second metal ion (such as zinc). A back titration is necessary if the analyte reacts too slowly with EDTA or blocks the indicator. In addition, the EDTA complex from the second metal ions must be less stable than the corresponding analyte complex.

Example 6.6

Al^{3+} could be analyzed by a back titration, using standard Zn^{2+} at pH=5-6 with a xylenol orange indicator. A solution containing 20.00 mL of Al^{3+} in dilute HCl is treated with 25.00 mL of 0.052 80 $mol\cdot L^{-1}$ EDTA. Titration with 0.023 00 $mol\cdot L^{-1}$ Zn^{2+} requires 17.60 mL to reach the endpoint. What is the concentration of Al^{3+}?

Solution:

The unknown was treated with 25.00 mL of 0.052 80 mol · L^{-1} EDTA, which contains (25.00 mL × 0.052 80 mol · L^{-1}) = 1.320 mmol of EDTA. Back titration required (17.60 mL × 0.023 00 mol · L^{-1}) = 0.404 8 mmol of Zn^{2+}. Because 1 mol of EDTA reacts with 1 mol of metal ion, there must have been:

$$1.320 \text{ mmol EDTA} - 0.404\,8 \text{ mmol } Zn^{2+} = 0.915\,2 \text{ mmol } Al^{3+}$$

The concentration of Al^{3+} is 0.915 2 mmol/20.00 mL = 0.045 76 mol · L^{-1}

6.5.3 Displacement Titration

A **displacement titration** may be feasible for metal ions without an available indicator. In displacement titration, an unmeasured excess of the solution containing the Mg^{2+}-EDTA or Zn^{2+}-EDTA complex is added to the analyte (M^{n+}). After the reaction, M^{n+} would displace Mg^{2+} or Zn^{2+} from the metal-EDTA complex. The analyte forms a more stable complex (MY):

$$M^{n+} + MgY \longrightarrow MY + Mg^{2+}$$

6.5.4 Indirect Titration

Anions such as CO_3^{2-}, CrO_4^{2-}, S^{2-}, and SO_4^{2-} could not coordinate with EDTA, while cations such as Na or K could not form a stable EDTA complex. These ions can be analyzed with the EDTA by **indirect titration.**

For example, sulfate could be precipitated with excess Ba^{2+} at pH 1. The $BaSO_4$ is washed and then reacted with a known excess EDTA at pH 10 to bring Ba^{2+} back into the Ba^{2+}-EDTA complex. Finally, the excess EDTA is back-titrated with Mg^{2+}, taking Eriochrome Black T as an indicator.

Key Glossary

Complexometric titration: Complexometric titration is the titration analysis method based on the reaction of coordination compounds. Complexometric titration is EDTA titration, which is often used for quantitative analysis of metal ions, and the coordination ratio is often 1 : 1.

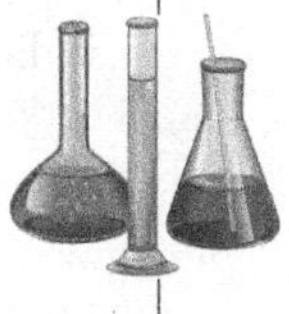

The formation constant: The constant of metal ion (M) and ligands (L) formed a coordination compound at a certain temperature, represented as K_f. The greater the value of K_f, the more stable the complexes.

The stepwise formation constant and **cumulative formation constant:** The formation of ML_n is step by step, and the stepwise formation constant is K_{f_n}. Multiply the K_{f_n} successively, the cumulative constants (β_n) can be obtained.

Side reaction coefficient and **conditional formation constant**: The side reaction coefficient measures the degree of influence from the side reaction to the main reaction. The acidic effect coefficient ($\alpha_{Y(H)}$) and coordination effect coefficient ($\alpha_{M(L)}$) are the main side reaction coefficients. The larger the value, the more serious the influence on the main reaction. The conditional formation constant K'_{MY} is the actual stability constant of EDTA-metal complexes, shown as $\lg K'_{MY} = \lg K_{MY} - \lg \alpha_{M(L)} - \lg \alpha_{Y(H)}$.

Metal ion indicator: A metal ion indicator can react with some metal ions to produce a significantly different color complex to indicate the endpoint. Metal ion indicators are compounds whose color changes when they bind to a metal ion. Useful indicators must bind metal less strongly than EDTA, $K'_{MY}/K'_{MIn} \geqslant 10^2$ at least.

Blocking: A metal ion does not freely dissociate from an indicator. The metal ion is said to be blocked by the indicator due to $\lg K_{MIn} > \lg K_{MY}$. The indicator does not change its color.

Essential Exercises

Review Questions

1. What are the characteristics of EDTA-metal ion complexes?

2. Explain the following nouns: absolute formation constant, stepwise formation constant, and cumulative formation constant.

3. What are the difference and relationship between the complexes' absolute formation constant and the conditional stability constant?

4. What is the acidic effect? What is the acidic effect coefficient?

5. Why is it necessary to control the solution's acidity in the complexometric titration?

6. What is a metal ion indicator? What is the principle of the metal ion indicator? What should it be? Please give an example to illustrate.

7. Why do we have a pH limit when we use a metal ion indicator? Why is the same indicator used to titrate different metal ions, but the pH conditions are different?

8. What is the ossification phenomenon of a metal ion indicator? How to avoid it?

9. At pH = 10.0, to determine Ca^{2+} with Eriochrome Black T, why add a small amount of MgY to the standard solution of EDTA?

10. Describe the function of the ETDA acid effect curve.

11. Determine the content of Pb^{2+} in a mixing solution including Bi^{3+}, Pb^{2+}, Al^{3+}, and Mg^{2+}, is there any interference of the other four metal ions?

12. In the complexometric titration analysis, how do we choose the titration condition when the co-existing ions exist?

13. Why should we add triethanolamine when determining Ca^{2+} and Mg^{2+} by complexometric titration? Is the pH in the specific operation first or first triethanolamine? Why?

14. What are the methods of complexometric titration? How do you apply these methods?

15. The content of Al^{3+} is determined by back titration; an overdose EDTA is first added to the solution at a pH of 3.5, and the Al^{3+} is heated to react sufficiently. Try to explain the reason for choosing this pH.

Cumulative Problems

1. To determine the content of each component in the following mixed solution using the complexometric titration (including titration ligands, acidity, indicator, and titration techniques).

(1) Zn^{2+}, Mg^{2+} mixed solution.

(2) Fe^{3+}, Cu^{2+}, Ni^{2+} mixed solution.

2. In EDTA titration, which statement about the acid effect is correct(   ).

A. The greater the pH, the greater the acid effect coefficient.

B. The greater the acid effect coefficient, the greater the stability of the complexes.

C. The smaller the acid effect coefficient, the greater the stability of the complexes.

D. The larger the acid effect coefficient, the larger the titration jump of the titration curve.

3. In EDTA-titration, the K'_{MIn} is required (   )

A. $K'_{MIn} > K'_{MY}$ B. $K'_{MIn} < K'_{MY}$

C. $K'_{MIn} = K'_{MY}$ D. $K'_{MY} > 100\ K'_{MIn}$

4. The color of the titration endpoint is(   )

A. The color of the free metal indicator.

B. The color of EDTA-metal ions complexes.

C. The color of EDTA-metal indicator complexes.

D. The mixed color of A and C.

5. There are four metal ions solutions of Mn^{2+}, Cd^{2+}, Fe^{3+}, and Cu^{2+}, whose concentrations are 1.0×10^{-2} mol · L^{-1} at pH=4.5. Use EDTA standard solution to titrate the four metal ions. If the acidic effect is considered only ($\lg \alpha_{Y(H)} = 7.4$), the metal ion that cannot be accurately titrated is(   ). Known as: $\lg K_{MnY}=7.4$, $\lg K_{CdY}=16.5$, $\lg K_{CdY}=16.5$, $\lg K_{FeY}=25.1$, $\lg K_{CuY}=18.8$.

A. Mn^{2+} B. Cd^{2+} C. Fe^{3+} D. Cu^{2+}

6. Titration of metal ions by EDTA, to achieve $E \leqslant 0.2\%$ at the endpoint, the conditions should be satisfied(   ).

A. $c\times K_{MY} \geqslant 10^{-8}$ B. $c\times K'_{MY} \geqslant 10^{6}$

C. $c\times K_{MY} \geqslant 10^{6}$ D. $\lg c\times K_{MY} \geqslant 6$

7. In EDTA titration, pH and acid effect coefficient $\alpha_{Y(H)}$ influence the coordination titration(   ).

A. pH increases, $\alpha_{Y(H)}$ increases, and the stability of the complexes increases.

B. pH increases, $\alpha_{Y(H)}$ decreases, and the stability of the complexes increases.

C. The acidity increases, $\alpha_{Y(H)}$ increases, and the stability of the complexes increases.

D. The acidity increases, $\alpha_{Y(H)}$ decreases, and the pH range of titration jumps larger.

8. The discriminant for accurate titration by EDTA is(   ); factors affecting the range of titration jump are (   ).

9. When Ca^{2+} and Mg^{2+} ions coexist in the solution, the method of each content is determined respectively: pH = (   ), determine (   ) by EDTA standard solution; take the same volume solution and add (   ) to make Mg^{2+} becoming (   ), then titrated by EDTA standard solution.

10. At pH = 12, a calcium indicator is used to determine the content of CaO in limestone. After the sample of 0.408 6 g is determined in a 250 mL volumetric flask, 25.00 mL test liquid is absorbed by pipette, with EDTA titration, and 17.50 mL of EDTA solution of 0.020 40 mol · L^{-1} is used to find the percentage of CaO in the limestone.

11. The solution of EDTA was determined by 25.00 mL Ca^{2+} standard solution (including CaO 1.00 mg · mL^{-1}), 24.30 mL EDTA was used. If this EDTA solution titrates magnesium salt, the Eriochrome Black T is the indicator, titrated at pH = 10, and consumed EDTA 36.74 mL. How much is MgO contained in magnesium salt?

12. Weighed phosphorus sample for 0.100 0 g and dissolved. Made the phosphorus to $MgNH_4PO_4$ as precipitation, after washing, dissolved again, titrated by 0.010 0 mol · L^{-1} EDTA standard solution of 20.00 mL, calculate the percentage of P_2O_5 in the sample.

13. Measured the total hardness of water at pH = 10, choosing Eriochrome Black T as the indicator. A sample of 100.0 mL water is reacted by 18.90 mL 0.010 00 mol · L^{-1}. For the EDTA standard solution, how much are the Germany degrees of hardness of the sample? Take the same water sample for 100.0 mL, pH = 12, titrated by 12.62 mL 0.010 00 mol · L^{-1} EDTA standard solution with a calcium indicator, and how much should be found in each liter of water containing Ca^{2+} and Mg^{2+}? (10 mg CaO per liter is one Germany degree).

Chapter 7

Redox Titration

Key Points

- *Characteristic of Redox Titration*
- *Electrode Potential*
- *Degree of Redox Reaction*
- *Factors Influencing Redox Reaction Rate*
- *Titration Curves and Indicator of Redox Titration*
- *Potassium Permanganate Titration*
- *Potassium Dichromate Titration*
- *Iodimetry Titration*

7. 1 Characteristics of Redox Titration

Oxidation-reduction(redox) titrations are similar in setup to acid-base titrations, but the focus is different. On the other hand, acid-base titrations follow the movement of protons. Redox titrations follow the transfer of charge (as electrons) to reach the equivalence point. Redox titrations can utilize indicators that change color at a particular voltage value (φ).

A redox titration is based on an oxidation-reduction reaction between analytic and titrant. In addition to the many common chemistry, biology, and environmental and materials science analyses that can be measured by redox titrations, exotic oxidation states of elements in uncommon materials such as superconductors and laser materials are measured by redox titrations. For example, chromium added to laser crystals to increase efficiency is found in the common oxidation states +3 and +6, and the unusual +4. A redox titration is an excellent way to unravel the nature of this complex mixture of chromium ions. The redox titrations have some different characters from other titrations, such as:

(1)The mechanism of redox titration is more complex, and there is more side-reacting, so the problem related to the stoichiometry is more complex.

(2)The redox reaction is slower than the other types of reaction.

(3) There are more redox titration methods than other types of titration because oxidants and reductants can all be used as titrants.

(4)Redox titration cannot only be used to detect the redox quality substance but also to detect other non-redox quality substances.

7.2 Electrode Potential

Electrode potential is a measure of the driving force of the electrode reaction and is temperature and activity-dependent, and appears at the interface between an electrode and electrolyte due to the transfer of charged species across the interface, specific adsorption of ions at the interface, and specific adsorption or orientation of polar molecules, including those of the solvent. Electrode potential is the electric potential on an electrode component. In a cell, there is an electrode potential for the cathode and an electrode potential for the anode. Suppose a solution forms part of an electrochemical cell. In that case, the cell's potential, the current flowing through it, and its resistance are all determined by the chemical composition of the solution. Quantitative and qualitative information can thus be obtained by measuring one or more of these electrical properties under controlled conditions. Direct measurements can be made in which sample solutions are compared with standards; alternatively, the changes in an electrical property during titration can be followed to detect the equivalence point.

7.2.1 Standard Electrode Potential

The oxidation abilities of various oxidants and reduction abilities of various oxidants are different. The redox capacity of compounds can be measured by electrode potential. For any reversible redox pair:

$$\text{Ox(oxidation state)} + n\text{e}^- \rightleftharpoons \text{Red(reductive state)}$$

The Nernst equation shows the relationship between such activities and the electrode potential.

$$\varphi_{\text{Ox/Red}} = \varphi^{\ominus}_{\text{Ox/Red}} + \frac{RT}{nF}\ln\frac{a_{\text{Ox}}}{a_{\text{Red}}} \qquad \textbf{(Equation 7.1)}$$

Where φ is the electrode potential, $\varphi^{\ominus}$ is the standard potential, a_{Ox} and a_{Red} are the activities of the oxidation state and reductive state, respectively, R is the gas constant, T is the thermodynamic temperature, F is the Faraday constant (96 485 $\text{C} \cdot \text{mol}^{-1}$) and n is the number of electrons involved in the electrode reaction. At 298.15 K, the equation is of the form:

$$\varphi_{\text{Ox/Red}} = \varphi^{\ominus}_{\text{Ox/Red}} + \frac{0.059\,2}{n}\lg\frac{a_{\text{Ox}}}{a_{\text{Red}}} \qquad \textbf{(Equation 7.2)}$$

Standard electrode potential is the measure of the individual potential of a reversible electrode at standard state, which is with solutes at an effective concentration of $a_{\text{Ox}} = a_{\text{Red}} = 1\ \text{mol} \cdot \text{L}^{-1}$, At 298.15 K, and gases at a pressure of 101.325 kPa.

7.2.2 Condition Electrode Potential

If the concentration replaces the activity, the activity coefficient should be quoted. γ_{Ox} and γ_{Red}.

$$a_{Ox}=\gamma_{Ox}[Ox] \qquad a_{Red}=\gamma_{Red}[Red]$$

In addition, some side reactions will occur in the oxidation state. The reduced state when the solution medium is different, such as acid-base reaction, precipitation reaction, coordination reaction, and so on, can affect the electrode potential, so it is necessary to consider the occurrence of these side reactions and introduce the side reaction coefficient β_{Ox} and β_{Red}.

$$a_{Ox}=\gamma_{Ox}[Ox]=\gamma_{Ox}\frac{c_{Ox}}{\beta_{Ox}} \qquad a_{Red}=\gamma_{Red}[Red]=\gamma_{Red}\frac{c_{Red}}{\beta_{Red}}$$

Therefore, the Nernst equation will be:

$$\varphi_{Ox/Red}=\varphi^{\ominus}_{Ox/Red}+\frac{0.059\ 2}{n}\lg\frac{\gamma_{Ox}\beta_{Red}c_{Ox}}{\gamma_{Red}\beta_{Ox}c_{Red}}$$

When $c_{Ox}=c_{Red}=1\ mol\cdot L^{-1}$, the Nernst equation will be:

$$\varphi^{\ominus'}_{Ox/Red}=\varphi^{\ominus}_{Ox/Red}+\frac{0.059\ 2}{n}\lg\frac{\gamma_{Ox}\beta_{Red}}{\gamma_{Red}\beta_{Ox}} \qquad \textbf{(Equation 7.3)}$$

$\varphi^{\ominus'}_{Ox/Red}$ is named the **condition electrode potential**, which means the electrode potential under special conditions of $c_{Ox}=c_{Red}=1\ mol\cdot L^{-1}$.

Therefore, the Nernst equation will be:

$$\varphi_{Ox/Red}=\varphi^{\ominus'}_{Ox/Red}+\frac{0.059\ 2}{n}\lg\frac{c_{Ox}}{c_{Red}} \qquad \textbf{(Equation 7.4)}$$

Example 7.1

Known about $\varphi^{\ominus}(Fe^{3+}/Fe^{2+})=0.771$ V, calculate the electrode potential of Fe^{3+}/Fe^{2+} at conditions of $c(Fe^{3+})=1.0\ mol\cdot L^{-1}$, $c(Fe^{2+})=0.000\ 1\ mol\cdot L^{-1}$.

Solution:

$$\varphi(Fe^{3+}/Fe^{2+})=\varphi^{\ominus}(Fe^{3+}/Fe^{2+})+\frac{0.059\ 2}{1}\lg\frac{[Fe^{3+}]}{[Fe^{2+}]}$$

$$\varphi(Fe^{3+}/Fe^{2+})=0.771+0.059\ 2\lg\frac{1.0}{0.000\ 1}=1.0(V)$$

Example 7.2

Calculate the electrode potential of Ce^{4+}/Ce^{3+} at following conditions of solution, $c(HCl)=1.0\ mol \cdot L^{-1}$, $c(Ce^{4+})=0.01\ mol \cdot L^{-1}$, $c(Ce^{3+})=0.001\ mol \cdot L^{-1}$.

Solution:

The condition electrode potential $\varphi^{\ominus'}(Ce^{4+}/Ce^{3+})=1.28$ V can be checked in the appendix table.

$$\varphi(Ce^{4+}/Ce^{3+})=\varphi^{\ominus'}(Ce^{4+}/Ce^{3+})+\frac{0.0592}{1}\lg\frac{[Ce^{4+}]}{[Ce^{3+}]}$$

$$\varphi(Ce^{4+}/Ce^{3+})=1.28+0.0592\lg\frac{0.01}{0.001}=1.34(V)$$

If the influence of the medium is not considered, the $\varphi^{\ominus}(Ce^{4+}/Ce^{3+})$ is used for calculation rather then $\varphi^{\ominus'}(Ce^{4+}/Ce^{3+})$. $\varphi^{\ominus}(Ce^{4+}/Ce^{3+})=1.72$ V checked in the appendix table.

$$\varphi(Ce^{4+}/Ce^{3+})=\varphi^{\ominus}(Ce^{4+}/Ce^{3+})+\frac{0.0592}{1}\lg\frac{[Ce^{4+}]}{[Ce^{3+}]}$$

$$\varphi(Ce^{4+}/Ce^{3+})=1.72+0.0592\lg\frac{0.01}{0.001}=1.78(V)$$

The results show that the electrode potential can be affected dramatically by the situation of the medium.

7.2.3 Factors Affecting the Conditions Electrode Potential

The factors affecting the conditions of electrode potential are main ionic strength and side reactions, including coordination reaction, precipitation reaction, and acid effects.

(1) Precipitation affects the conditional electrode potential If the oxidation state or reduction state does a precipitation side reaction, the electrode potential would be changed. Precipitating for oxidation state can decrease the electrode potential, and precipitating for the reduction state can increase the electrode potential.

Example 7.3

Calculate the $\varphi(Cu^{2+}/Cu)$ at the condition of $c(CuI)=1.0\ mol \cdot L^{-1}$, known about $\varphi^{\ominus}(Cu^{2+}/Cu)=0.17$ V, $K_{sp}^{\ominus}=1.1\times10^{-12}$.

Solution:

$$\varphi(Cu^{2+}/Cu)=\varphi^{\ominus}(Cu^{2+}/Cu)+\frac{0.0592}{1}\lg\frac{[Cu^{2+}]}{[Cu^{+}]}$$

$$=\varphi^{\ominus}(Cu^{2+}/Cu)+0.0592\lg\frac{[Cu^{2+}]\times[I^-]}{K^{\ominus}_{sp}(CuI)}$$

$$=0.17+0.0592\lg\frac{1}{1.1\times10^{-12}}$$

$$=0.84(V)$$

(2) Complexation affects the conditional electrode potential There are usually various anions with complexing functions in solution that can possibly coordinate with the oxidation or reduction states, which will lead the electrode potential to vary.

Example 7.4

Calculate the $\varphi(Fe^{3+}/Fe^{2+})$ in solution of $c(Fe^{3+})=0.1\ mol\cdot L^{-1}$, $c(Fe^{2+})=1.0\times10^{-5}\ mol\cdot L^{-1}$, $c(F^-)=1.0\ mol\cdot L^{-1}$.

Solution:

The solution, known about the cumulative stability constants of complex $[FeF_3]$:

$$\beta_1=1.9\times10^5, \beta_2=2.0\times10^9, \beta_3=1.2\times10^{12},$$

$$\beta(FeF_3)=1+\beta_1c(F^-)+\beta_2c^2(F^-)+\beta_3c^3(F^-)\approx1.2\times10^{12}$$

$$c'(Fe^{3+})=\frac{c(Fe^{3+})}{\beta(FeF_3)}=\frac{0.1}{1.2\times10^{12}}=8.3\times10^{-14}(mol\cdot L^{-1})$$

$$\varphi(Fe^{3+}/Fe^{2+})=\varphi^{\ominus}(Fe^{3+}/Fe^{2+})+\frac{0.0592}{1}\lg\frac{[Fe^{3+}]}{[Fe^{2+}]}=0.83(V)$$

(3) Acidity affects the conditional electrode potential The electrode potential will also be changed to account for H^+ or OH^-, participating in the redox reaction.

Example 7.5

At 298.15 K, the electrode reaction:

$$Cr_2O_7^{2-}(aq)+14H^+(aq)+6e^-\rightleftharpoons2Cr^{3+}(aq)+7H_2O$$

Known about: $\varphi^{\ominus}(Cr_2O_7^{2-}/Cr^{3+})=1.33\ V$, $c(Cr_2O_7^{2-})=1.0\ mol\cdot L^{-1}$, $c(Cr^{3+})=1.0\ mol\cdot L^{-1}$. Calculate $\varphi(Cr_2O_7^{2-}/Cr^{3+})=?$ At following conditions:

(A) $c(H^+)=10.0\ mol\cdot L^{-1}$

(B) $c(H^+)=10^{-4}\ mol\cdot L^{-1}$

Solution:

(A) $\varphi(Cr_2O_7^{2-}/Cr^{3+})=\varphi^{\ominus}(Cr_2O_7^{2-}/Cr^{3+})-\frac{0.0592\ V}{n}\times\lg\frac{[Cr^{3+}]^2}{[Cr_2O_7^{2-}][H^+]^{14}}$

$=1.33\ V-\frac{0.0592\ V}{6}\times\lg\frac{1.0}{1.0\times10^{14}}$

$=1.33\ V+0.14\ V$

$=1.47\ V$

(B) $\varphi(Cr_2O_7^{2-}/Cr^{3+})=\varphi^{\ominus}(Cr_2O_7^{2-}/Cr^{3+})-\frac{0.0592\ V}{n}\times\lg\frac{[Cr^{3+}]^2}{[Cr_2O_7^{2-}][H^+]^{14}}$

$=1.33\ V-\frac{0.0592\ V}{6}\times\lg\frac{1.0}{1.0\times(10^{-4})^{14}}$

$=1.33\ V-0.55\ V$

$=0.78\ V$

7.3 Degree of Redox Reaction

Redox titration requires the redox reaction to proceed thoroughly. The equilibrium constant usually measures the extent of the reaction. According to the Nernst equation, the equilibrium constant can be calculated with the conditional or standard electrode potential.

$$n_2Ox_1+n_1Red_2 \rightleftharpoons n_2Red_1+n_1Ox_2$$

The electrode potential of the half-reaction is, respectively:

$$\varphi_1=\varphi_1^{\ominus'}+\frac{0.0592}{n_1}\lg\frac{c_{Ox_1}}{c_{Red_1}};\varphi_2=\varphi_2^{\ominus'}+\frac{0.0592}{n_2}\lg\frac{c_{Ox_2}}{c_{Red_2}}$$

The electrode potentials of the half-reaction are equal to when the redox reaction reaches balance

$$\varphi_1^{\ominus'}+\frac{0.0592}{n_1}\lg\frac{c_{Ox_1}}{c_{Red_1}}=\varphi_2^{\ominus'}+\frac{0.0592}{n_2}\lg\frac{c_{Ox_2}}{c_{Red_2}}$$

$$\varphi_1^{\ominus'}-\varphi_2^{\ominus'}=\frac{0.0592}{n_1n_2}\lg\frac{c_{Ox_2}^{n_1}c_{Red_1}^{n_2}}{c_{Red_2}^{n_1}c_{Ox_1}^{n_2}}=\frac{0.0592}{n_1n_2}\lg K'$$

$$\lg K'=\frac{n_1n_2(\varphi_1^{\ominus'}-\varphi_2^{\ominus'})}{0.0592} \quad \textbf{(Equation 7.5)}$$

$n_1n_2=n$, n is the electron number of transfer in the redox reaction.

$$\lg K'=\frac{n(\varphi_1^{\ominus'}-\varphi_2^{\ominus'})}{0.0592} \quad \textbf{(Equation 7.6)}$$

It can be found, according to Equation 7.5, that there is a positive relationship between K' and $(\varphi_1^{\ominus'}-\varphi_2^{\ominus'})$.

Example 7.6

Calculate the equilibrium constant of the reaction between Fe^{3+} and Sn^{2+} in the HCl medium, $c(HCl) = 1\ mol \cdot L^{-1}$. And judge whether the reaction can be quantitatively carried out.

Solution:

Known about $Fe^{3+} + e^{-} = Fe^{2+}$ $\quad \varphi^{\ominus'}(Fe^{3+}/Fe^{2+}) = 0.77\ V$

$$Sn^{4+} + 2e^{-} = Sn^{2+} \qquad \varphi^{\ominus'}(Sn^{4+}/Sn^{2+}) = 0.14\ V$$

Chemical reaction equation $2Fe^{3+} + Sn^{2+} \rightleftharpoons Sn^{4+} + 2Fe^{2+}$

$$\lg K' = \frac{n(\varphi_1^{\ominus'} - \varphi_2^{\ominus'})}{0.059\ 2} = \frac{n(0.77 - 0.14)}{0.059\ 2} = 21.28$$

7.4 Factors Influencing Redox Reaction Rate

The main factors affecting the rate of redox reaction are the following aspects: properties of oxidants and reducers, concentration of reactant, use of catalyst, temperature, and induced reaction. Most of these factors are discussed in general chemistry, so that we will concentrate on the influence of induced reactions.

The inductive effect is a method to accelerate a slower chemical reaction through another reaction, where one chemical reaction is naturally slow. However, when adding a small amount of other compounds into the reaction system, which will react with the reactant, it will be induced to accelerate the reaction rate. Such as, the reaction rate of the redox reaction between potassium permanganate ($KMnO_4$) and potassium chloride (KCl) is extremely slow, but it can be accelerated by the other reaction occurring between $KMnO_4$ and ferrous ion (Fe^{2+}) added.

7.5 Titration Curves and Indicator of Redox Titration

7.5.1 Titration Curves of Redox Titration

In sections on acid-base titration, it is shown how the change in pH during acid-base titrations may be calculated and how the titration curves thus obtained can be used to ascertain the most suitable indicator to be used in a given titration and to determine the titration error. Similar procedures may be carried out for oxidation-reduction titrations. Consider first a simple case that involves only a change in ionic charge and is theoretically independent of the hydrogen-ion concentration.

1. Potential calculation at the stoichiometric point

Redox reaction $n_1 Red_2 + n_2 Ox_1 \rightleftharpoons n_1 Ox_2 + n_2 Red_1$

Half-reaction $Ox_1 + n_1 e^- \rightleftharpoons Red_1$ $\varphi_1 = \varphi_1^{\ominus'} + \frac{0.0592}{n_1}\lg\frac{c_{Ox_1}}{c_{Red_1}}$ (1)

$Ox_2 + n_2 e^- \rightleftharpoons Red_2$ $\varphi_2 = \varphi_2^{\ominus'} + \frac{0.0592}{n_2}\lg\frac{c_{Ox_2}}{c_{Red_2}}$ (2)

At the stoichiometric point $\varphi_{sp} = \varphi_1 = \varphi_2$, add up the two formulas after doing with $(1)\times n_1 + (2)\times n_2$.

$$n_1\varphi_1 + n_2\varphi_2 = n_1\varphi_1^{\ominus'} + n_2\varphi_2^{\ominus'} + 0.0592\lg\frac{c_{Ox_1}}{c_{Red_1}} + 0.0592\lg\frac{c_{Ox_2}}{c_{Red_2}}$$

$$(n_1+n_2)\varphi_{sp} = n_1\varphi_1^{\ominus'} + n_2\varphi_2^{\ominus'} + 0.0592\lg\left(\frac{c_{Ox_1}}{c_{Red_1}}\times\frac{c_{Ox_2}}{c_{Red_2}}\right)$$

At the stoichiometric point $\frac{c_{Ox_1}}{c_{Red_2}} = \frac{n_1}{n_2}$; $\frac{c_{Ox_2}}{c_{Red_1}} = \frac{n_2}{n_1}$, therefore $\lg\left(\frac{c_{Ox_1}}{c_{Red_1}}\times\frac{c_{Ox_2}}{c_{Red_2}}\right) = \lg 1 = 0$

$$(n_1+n_2)\varphi_{sp} = n_1\varphi_1^{\ominus'} + n_2\varphi_2^{\ominus'}$$

$$\varphi_{sp} = \frac{n_1\varphi_1^{\ominus'} + n_2\varphi_2^{\ominus'}}{n_1+n_2} \qquad \textbf{(Equation 7.7)}$$

If $n_1 = n_2$

$$\varphi_{sp} = \frac{\varphi_1^{\ominus'} + \varphi_2^{\ominus'}}{2} \qquad \textbf{(Equation 7.8)}$$

2. Calculation of titration jump

(1) Potential before the stoichiometric point It can be evaluated by the titrated object here Red_2 assumed as the titrated object:

$$\varphi_{Ox_2/Red_2} = \varphi_{Ox_2/Red_2}^{\ominus'} + \frac{0.0592}{n_2}\lg\frac{c_{Ox_2}}{c_{Red_2}} \qquad \textbf{(Equation 7.9)}$$

(2) Potential behind stoichiometric point It can be evaluated through titrant here Red_2 assumed as titrant:

$$\varphi_{Ox_1/Red_1} = \varphi_{Ox_1/Red_1}^{\ominus'} + \frac{0.0592}{n_1}\lg\frac{c_{Ox_1}}{c_{Red_1}} \qquad \textbf{(Equation 7.10)}$$

3. Example

A suitable example, for purposes of illustration, is the titration of 100 mL of 0.1 mol • L^{-1} iron (Ⅱ) with 0.1 mol • L^{-1} cerium (Ⅳ) in the presence of dilute sulphuric acid:

$$Ce^{4+} + Fe^{2+} \rightleftharpoons Ce^{3+} + Fe^{3+}$$

Changes in solution composition during titration are shown in the following.

Titration process	Solution composition
Before the start of the titration	Fe^{2+}
Stoichiometric point	Fe^{2+}, Fe^{3+}, Ce^{3+}, Ce^{4+} (content is very small)
Behind stoichiometric point	Fe^{3+}, Ce^{3+}, Ce^{4+}, Fe^{2+} (content is very small)

The quantity corresponding to $[H^+]$ in acid-base titrations is the ratio [Ox]/[Red]. We are concerned with two systems: the $[Fe^{3+}]/[Fe^{2+}]$ ion electrode and the $[Ce^{4+}]/[Ce^{3+}]$ ion electrode.

$$\varphi_1=\varphi_1^{\ominus}+\frac{0.059\ 2}{1}\log\frac{[Fe^{3+}]}{[Fe^{2+}]}=+0.75+0.059\ 2\log\frac{[Fe^{3+}]}{[Fe^{2+}]}$$

$$\varphi_2=\varphi_2^{\ominus}+\frac{0.059\ 2}{1}\log\frac{[Ce^{4+}]}{[Ce^{3+}]}=+1.45+0.059\ 2\log\frac{[Ce^{4+}]}{[Ce^{3+}]}$$

The equilibrium constant of the reaction is given by:

$$\log K=\log\frac{[Ce^{3+}]\times[Fe^{3+}]}{[Ce^{4+}]\times[Fe^{2+}]}=\frac{1}{0.059\ 2}(1.45-0.75)=11.82$$

or

$$K=7\times10^{11}$$

The reaction is, therefore, virtually complete.

During the addition of the cerium (Ⅳ) solution up to the equivalence point, its only effect will be to oxidize the iron (Ⅱ) (since K is large) and consequently change the ratio $[Fe^{3+}]/[Fe^{2+}]$. When 10 mL of the oxidizing agent has been added, $[Fe^{3+}]/[Fe^{2+}]$ = 10/90 (approx.) and

$$\varphi_1=0.75+0.059\ 2\ \lg\frac{10}{90}=0.75-0.056=0.69(V)$$

With 50 mL of the oxidizing agent, $\varphi_1=\varphi_1^{\ominus}=0.75(V)$

With 90 mL, $\varphi_1=0.75+0.059\ 2\ \lg\frac{90}{10}=0.81(V)$

With 99 mL, $\varphi_1=0.75+0.059\ 2\ \lg\frac{99}{1}=0.87(V)$

With 99.9 mL, $\varphi_1=0.75+0.059\ 2\ \lg\frac{99.9}{0.1}=0.93(V)$

At the equivalence point (100.0 mL) $[Fe^{3+}]=[Ce^{3+}]$ and $[Ce^{4+}]=[Fe^{2+}]$

The electrode potential is given by:

$$\frac{\varphi_1^{\ominus}+\varphi_2^{\ominus}}{2}=\frac{0.75+1.45}{2}=1.10(V)$$

The subsequent addition of cerium (Ⅳ) solution will merely increase the ratio $[Ce^{4+}]/$

$[Ce^{3+}]$. Thus:

With 100.1 mL, $\varphi_2 = 1.45 + 0.0592 \lg \frac{0.1}{100} = 1.27(V)$

With 101 mL, $\varphi_2 = 1.45 + 0.0592 \lg \frac{1}{100} = 1.33(V)$

With 110 mL, $\varphi_2 = 1.45 + 0.0592 \lg \frac{10}{100} = 1.39(V)$

With 190 mL, $\varphi_2 = 1.45 + 0.0592 \lg \frac{90}{100} = 1.45(V)$

These results are shown in Table 7.1.

Table 7.1 The calculation results of electrode potential during the titration

$V(Ce^{4+})$/mL	Fe^{2+} percentage titrated/%	φ/V
0.00	5.0	0.60
2.00	10.0	0.62
4.00	20.0	0.64
8.00	40.0	0.67
10.00	50.0	0.68
12.00	60.0	0.69
18.00	90.0	0.74
19.80	99.0	0.80
19.98	99.9	0.86
20.00	100.0	1.06
20.02	100.1	1.26
22.00	110.0	1.38
30.00	150.0	1.42
40.00	200.0	1.44

4. Titration curve

Drawing with the percentage of additional titrant as abscissa and electrode potential as ordinate (Figure 7.1).

5. Titration jump

Previous calculations show that the titration jump size depends on the potential difference and number of transferred electrons between two electrodes. The bigger the potential difference between two electrodes, the bigger the titration jump (Figure 7.2).

When in a redox reaction $n_1 = n_2 = 1$, the stoichiometric point lies in the middle of the titration jump, and the titration curve near the stoichiometric point is symmetrical.

When in a redox reaction $n_1 \neq n_2$, the stoichiometric point is not at the center of the titration jump but in the direction of the electric-pair side, which has more electron gain and loss.

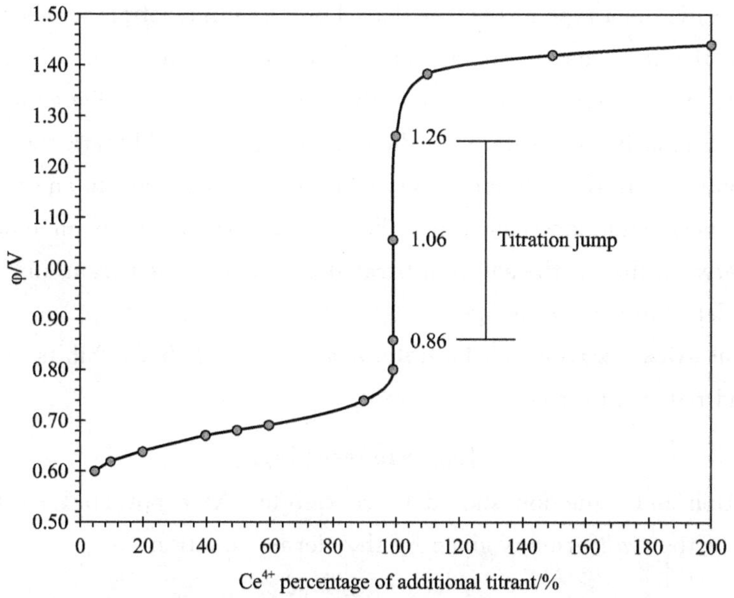

Figure 7.1 Titration curve of 20 mL of 0.1 mol · L^{-1} iron (Ⅱ) with 0.1 mol · L^{-1} cerium (Ⅳ) sulphate (calculated)

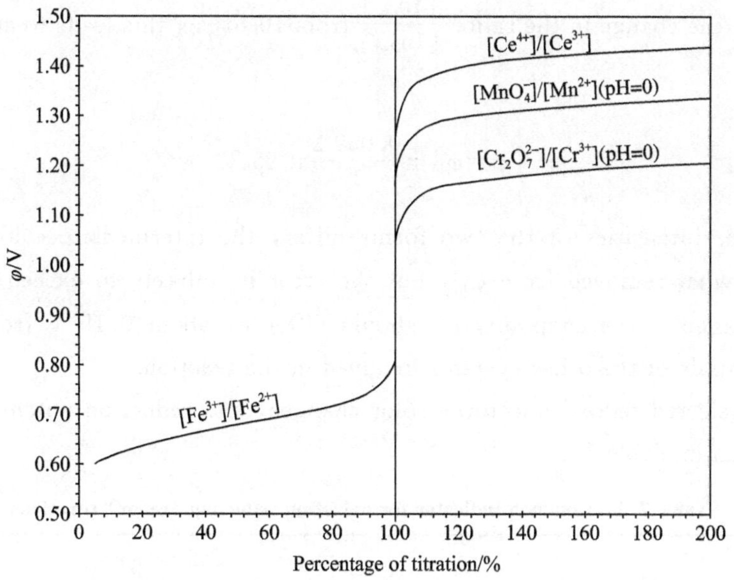

Figure 7.2 Titration curves of different oxidants (Ce^{4+}, MnO_4^- and $Cr_2O_7^{2-}$) react with reductant (Fe^{2+})

7.5.2 Indicator of Redox Titration

A redox titration is a type of titration based on a redox reaction between the analyst and titrant. Redox titration may involve using a redox indicator and a potentiometer. As discussed in sections of acid-base indicators, they mark the sudden change in pH during acid-base titrations. Similarly, an oxidation-reduction indicator should mark the sudden change in the oxidation potential in the neighborhood of the equivalence point in an oxidation-reduction titration. The ideal oxidation-reduction indicator will be one with an oxidation potential intermediate between that of the solution titrated and that of the titrant and which exhibits a sharp, readily detectable color change.

(1) **Internal oxidation-reduction indicators**, a compound that exhibits different colors in the oxidized and reduced forms:

$$In_{Ox} + ne^- \rightleftharpoons In_{Red}$$

The oxidation and reduction should be reversible. At a potential $\boldsymbol{E}$, the ratio of the concentrations of the two forms is given by the Nernst equation:

$$\varphi = \varphi_{In}^{\ominus} + \frac{RT}{nF}\lg\frac{[In_{Ox}]}{[In_{Red}]}$$

Where $\varphi^{\ominus}$ is the standard (strictly the formal) potential of the indicator. If the color intensities of the two forms are comparable, a practical estimate of the color-change interval corresponds to the change in the ratio. $\frac{[In_{Ox}]}{[In_{Red}]}$ from 10 to $\frac{1}{10}$, this leads to an interval of the potential of:

$$\varphi = \varphi_{In}^{\ominus} \pm \frac{0.059\ 2}{1} \text{ at } 25\ ℃$$

If the color intensities of the two forms differ, the intermediate color is attained at potential somewhat removed from $\varphi_{In}^{\ominus}$, but the error is unlikely to exceed 0.06 V. For a sharp color change at the endpoint, $\varphi_{In}^{\ominus}$ should differ by about 0.15 V from the standard (formal) potentials of the other systems involved in the reaction.

A list of selected redox indicators, color changes, and reduction potentials in an acidic medium is given in Table 7.2.

Table 7.2 Common indicator for oxidation-reduction (redox) titrations

Indicator	Color change		Potential pH=0/V
	Ox-form	Red-form	
5-Nitro-1,10-phenanthroline iron(Ⅱ) sulphate (nitroferroin)	Pale blue	Red	1.25
1,10-Phenanthroline iron(Ⅱ) sulphate (ferroin)	Pale blue	Red	1.06

Table 7.2 continued

Indicator	Color change		Potential pH=0/V
	Ox-form	Red-form	
2′2′-Bipyridyl iron(Ⅱ) sulphate	Faint blue	Red	1.02
5,6-Dimethylferroin	Pale blue	Red	0.97
N-Phenylanthranilic acid	Purple red	Colorless	0.89
4,7-Dimethyl-1,10-phenanthrolineir on (Ⅱ) sulphate (4,7-dimethylferroin)	Pale blue	Red	0.88
Diphenylaminesulphonic acid	Red-violet	Colorless	0.85
Diphenylbenzidine	Violet	Colorless	0.76
Diphenylamine	Violet	Colorless	0.76
3,3′-Dimethylnaphthidine	Purplish-red	Colorless	0.71
Starch-I,KI	Blue	Colorless	0.53
Methylene blue	blue	Colorless	0.52

(2) **Self-indicating reagents** are well illustrated by potassium permanganate, one drop of which will impart a visible pink coloration to several hundred milliliters of solution, even in the presence of slightly colored ions, such as iron (Ⅲ). The colors of cerium (Ⅳ) sulfate and iodine solutions have also been employed in the detection of endpoints, but the color change is not so marked as for potassium permanganate; here, however, sensitive internal indicators [1,10-phenanthroline-iron (Ⅱ) ion or N-phenyl anthranilic acid and starch respectively] are available.

This method has the drawback that an excess of oxidizing agent is always present at the endpoint. For work of the highest accuracy, the indicator blank may be determined and allowed for, or the error may be considerably reduced by performing the standardization and determination under similar experimental conditions.

(3) **The specific indicator** is not redox but can react with an oxidant or reductant to produce a color reaction. Moreover, the color reaction is irreversible, so the titration endpoint can be indicated. The most common special indicator is starch, such as soluble starch, which reacts with iodine solution to form blue compounds. Therefore, starch is often used as an indicator in iodimetry.

7.6 Common Redox Titration

Titrimetric methods based on the direct use of redox reactions are widely used. Their application to determining metals with two well-defined oxidation states is well known. Among the most important indirect methods of analysis that employ redox reactions are the bromination procedures for determining aromatic amines, phenols, and other compounds that undergo stoichiometric bromine substitution or addition. Common redox titrations are potassium permanganate titration, potassium dichromate titration, and iodimetry titration.

7.6.1 Potassium Permanganate Titration

7.6.1.1 Overview

Potassium permanganate is an inorganic chemical compound and medication. As a medication, it is used to clean wounds and dermatitis. Almost all applications of potassium permanganate exploit its oxidizing properties. As a strong oxidant that does not generate toxic byproducts, $KMnO_4$ has many niche uses. In acid solutions, the reduction can be represented by the following equation:

$$MnO_4^- + 8H^+ + 5e^- \rightleftharpoons Mn^{2+} + 4H_2O \qquad \varphi^{\ominus} = 1.51\ V$$

The standard potential in acid solution, $\varphi^{\ominus}$ has been calculated to be 1.51 V; hence, the permanganate ion in an acid solution is a strong oxidizing agent.

Sulfuric acid is the most suitable acid, as it has no action upon permanganate in a dilute solution. With hydrochloric acid, there is a likelihood of the reaction:

$$2MnO_4^- + 10Cl^- + 16H^+ = 2Mn^{2+} + 5Cl_2 + 8H_2O$$

For the titration of colorless or slightly colored solutions, the use of an indicator is unnecessary since as little as 0.01 mL of 0.02 mol • L^{-1} potassium permanganate imparts a pale pink color to 100 mL of water. The intensity of the color in dilute solutions may be enhanced, if desired, by the addition of a redox indicator (such as sodium diphenylamine sulphonate, N-phenyl anthranilic acid, or ferroin) just before the endpoint of the reaction; this is usually not required but is advantageous if more dilute solutions of permanganate are used.

Potassium permanganate may also be used in strongly alkaline solutions. Here, two consecutive partial reactions take place.

①The relatively rapid reaction:

$$MnO_4^- + e^- \rightleftharpoons MnO_4^{2-} \qquad \varphi^{\ominus} = 0.564\ V$$

②The relatively slow reaction:

$$MnO_4^{2-} + 2H_2O + 2e^- \rightleftharpoons MnO_2 + 4OH^- \qquad \varphi^{\ominus} = 0.602\ V$$

The standard potential ($\varphi^{\ominus}$) of reaction ① is 0.564 V, and of reaction ② is 0.602 V. By suitably controlling the experimental conditions (e. g., adding barium ions, which form the sparingly soluble barium manganite as a fine, granular precipitate), reaction ① occurs almost exclusively. In moderately alkaline solutions, permanganate is reduced quantitatively to manganese dioxide. The half-cell reaction is:

$$MnO_4^- + 2H_2O + 3e^- \rightleftharpoons MnO_2 + 4OH^- \qquad \varphi^\ominus = 0.593\ V$$

And the standard potential $\varphi^\ominus = 0.593$ V.

7.6.1.2 Preparation of potassium permanganate standard solution

Potassium permanganate is not a primary standard. It is difficult to obtain a pure and completely free substance from manganese dioxide. Moreover, ordinary distilled water will likely contain reducing substances (traces of organic matter, etc.), which will react with the potassium permanganate to form manganese dioxide. The presence of the latter is very objectionable because it catalysis the auto-decomposition of the permanganate solution on standing:

$$MnO_4^- + 2H_2O = 4MnO_2 + 3O_2 + 4OH^-$$

Permanganate is inherently unstable in the presence of manganese (Ⅱ) ions:

$$MnO_4^- + 3Mn^{2+} + 2H_2O = 5MnO_2 + 2H^+$$

Potassium permanganate solutions may be standardized using arsenic (Ⅲ) oxide or sodium oxalate as primary standards.

7.6.1.3 Standardization of potassium permanganate solutions

This reagent is readily obtained pure and anhydrous, and the ordinary material has a purity of at least 99.9%. The experimental procedure originally employed a solution of the oxalate, acidified with dilute sulphuric acid and warmed to 80-90 ℃, which was titrated with the permanganate solution slowly (10-15 mL • min^{-1}) and with constant stirring until the first permanent faint pink color was obtained; the temperature near the endpoint was not allowed to fall below 60 ℃. However, with this procedure, the results may be 0.1%-0.45% high; the titer depends upon the acidity, the temperature, the rate of addition of the permanganate solution, and the stirring speed. Because of this, it is best to make a more rapid addition of 90%-95% of the permanganate solution (about 25-35 mL • min^{-1}) to a solution of sodium oxalate in 1 mol • L^{-1} sulphuric acid at 25-30 ℃, the solution is then warmed to 55-60 ℃, and the titration is completed, the last 0.5-1 mL portion being added dropwise. The method is accurate at 0.06 percent. Full experimental details are given below.

$$2Na^+ + C_2O_4^{2-} + 2H^+ \rightleftharpoons H_2C_2O_4 + 2Na^+$$

$$2MnO_4^- + 5H_2C_2O_4 + 6H^+ = 2Mn^{2+} + 10CO_2 + 8H_2O$$

For the standardization reaction to be carried out quantitatively and quickly, the calibration should pay attention to the following titration conditions.

(1) Temperature The suitable temperature for titration of the oxalic acid solution is 70-85 ℃, which should not exceed 90 ℃. Otherwise, oxalic acid will decompose, which will make the results increase.

$$H_2C_2O_4 \xrightarrow{\geq 90\ ℃} H_2O + CO_2\uparrow + CO\uparrow$$

(2) Acidity The optimum acidity of the solution should be kept within the range of 0.5-1 mol · L^{-1}. If the acidity is low, precipitation of MnO_2 will be produced. If the acidity is high, oxalic acid will decompose.

(3) Rate of addition of the permanganate solution The reaction between MnO_4^- and $C_2O_4^{2-}$ is slow usually at the beginning stage, but then it accelerates owing to the newly formed ion Mn^{2+} which has a catalytic effect. Therefore, when titrating, it is slower at the beginning and faster at the back.

(4) Endpoint of the titration Titration with potassium permanganate solution until the solution is pink, which does not fade after 30. It is the endpoint of titration. Use the following formula to calculate the results.

$$c\left(\frac{1}{5}KMnO_4\right)=\frac{m(Na_2C_2O_4)}{(V-V_0)\times M\left(\frac{1}{2}Na_2C_2O_4\right)\times 10^{-3}} \quad \textbf{(Equation 7.11)}$$

Where $m(Na_2C_2O_4)$ means the mass (g) of $Na_2C_2O_4$ weighed, V means the volume (mL) of $KMnO_4$ solution consumed by $Na_2C_2O_4$ in the process of titration. V_0 means the volume (mL) of $KMnO_4$ consumed in the blank test, $M\left(\frac{1}{2}Na_2C_2O_4\right)=67.00\ g \cdot mol^{-1}$.

Example 7.7

①To prepare a 1.5 L solution of $c\left(\frac{1}{5}KMnO_4\right)=0.2\ mol \cdot L^{-1}$, how many grams should be weighed?

②To prepare 1.0 L solution of $T\left(Fe^{2+}/\frac{1}{5}KMnO_4\right)=0.006\ 00\ g \cdot mL^{-1}$, how many grams should be weighed?

Solution:

①Known about $M(KMnO_4)=158.03\ g \cdot mol^{-1}$, $M(Fe)=55.85\ g \cdot mol^{-1}$

Because of $m(KMnO_4)=c\left(\frac{1}{5}KMnO_4\right)\times V(KMnO_4)\times M\left(\frac{1}{5}KMnO_4\right)$

Therefore $m(KMnO_4)=1.5\times 0.2\times \frac{1}{5}\times 158.03=9.5(g)$

②Reaction equation $MnO_4^- + 5Fe^{2+} + 8H^+ \xlongequal{} Mn^{2+} + 5Fe^{3+} + 4H_2O$

$$c\left(\frac{1}{5}KMnO_4\right)=\frac{T\times 1\ 000}{M_{Fe}}=\frac{0.006\ 00\times 1\ 000}{55.85}=0.107(mol \cdot L^{-1})$$

$$m(KMnO_4)=c\left(\frac{1}{5}KMnO_4\right)\times V(KMnO_4)\times M\left(\frac{1}{5}KMnO_4\right)$$

$$m(KMnO_4)=0.107\times 1\times \frac{1}{5}\times 158.03=3.4(g)$$

7.6.1.4 Application example of potassium permanganate titration

1. Analysis of hydrogen peroxide

Hydrogen peroxide is usually encountered in the form of an aqueous solution containing about 6%, 12%, or 30% hydrogen peroxide, and frequently referred to as "20-volume", "40-volume", and "100-volume" hydrogen peroxide, respectively; this terminology is based upon the volume of oxygen liberated when the solution is decomposed by boiling. Thus, 1 mL of "100-volume" hydrogen peroxide will yield 100 mL of oxygen when measured at standard temperature and pressure.

The following reaction occurs when potassium permanganate solution is added to hydrogen peroxide solution acidified with dilute sulphuric acid:

$$2MnO_4^- + 5H_2O_2 + 6H^+ = 2Mn^{2+} + 5O_2 + 8H_2O$$

Procedure: transfer 25.0 mL of the "20-volume" solution using a burette to a 500 mL graduated flask and dilute with water to the mark. Shake thoroughly. Pipette 25.0 mL of this solution to a conical flask, dilute with 200 mL water, add 20 mL dilute sulphuric acid (1 : 5), and titrate with standard 0.02 mol · L^{-1} potassium permanganate to the first permanent, faintly pink color. Repeat the titration; two consecutive determinations should agree within 0.1 mL.

2. Analysis of metallic peroxides

A metallic peroxide, such as sodium peroxide, can be analyzed similarly, provided that care is taken to avoid loss of oxygen during the dissolution of the peroxide. This may be done by working in a medium containing boric acid, converted to the relatively stable "perboric acid" upon adding the peroxide.

7.6.2 Potassium Dichromate Titration

7.6.2.1 Overview

Potassium dichromate is not a powerful oxidizing agent as potassium permanganate, but it has several advantages over the latter substance. It can be obtained pure, stable up to its fusion point, and therefore an excellent primary standard. Standard solutions of exactly known concentration can be prepared by weighing out the pure dry Salt and dissolving it in the proper volume of water. Furthermore, the aqueous solutions are stable indefinitely if adequately protected from evaporation. Potassium dichromate is used only in acid solution and is reduced rapidly to a green chromium (Ⅲ) salt at the ordinary temperature. It is not reduced by cold hydrochloric acid, provided the acid concentration does not exceed 1 or 2 mol · L^{-1}. Dichromate solutions are less easily reduced by organic matter than permanganate solutions and are also stable towards light. Potassium dichromate is therefore of particular value in the determination of iron in iron ores: the ore is usually dissolved in hydrochloric acid, the iron (Ⅲ) reduced to iron (Ⅱ), and the solution then titrated with standard dichromate solution.

$$Cr_2O_7^{2-}+6Fe^{2+}+14H^+ \rightleftharpoons 2Cr^{3+}+6Fe^{3+}+7H_2O$$

In acid solution, the reduction of potassium dichromate may be represented as:

$$Cr_2O_7^{2-}+14H^++6e^- \rightleftharpoons 2Cr^{3+}+7H_2O$$

The green color due to the Cr^{3+} ions formed by the reduction of potassium dichromate makes it impossible to ascertain the end-point of a dichromate titration by simple visual inspection of the solution, and so a redox indicator must be employed, which gives a strong and unmistakable color change; this procedure has rendered obsolete the external indicator method which was formerly widely used. Suitable indicators for use with dichromate titrations include N-phenyl anthranilic acid (0.1% solution in 0.005 mol · L^{-1} NaOH) and sodium diphenylamine sulphonate (0.2% aqueous solution); the latter must be used in the presence of phosphoric (V) acid.

7.6.2.2 Preparation of potassium dichromate standard solution

1. Direct preparation

Analytical grade potassium dichromate has a purity of not less than 99.9% and is satisfactory for most purposes. Powder finely about 6 g of the analytical grade material in a glass or agate mortar and heat for 30-60 min in an air oven at 140-150 ℃. Allow to cool in a closed vessel in a desiccator. Weigh out accurately about 5.88 g of the dry potassium dichromate into a weighing bottle and transfer the salt quantitatively to a 1 L graduated flask, using a small funnel to avoid loss. Dissolve the salt in the water flask and fill it to the mark; shake well. Alternatively, place a little over 5.88 g of potassium dichromate in a weighing bottle and weigh accurately. Empty the salt into a 1 L graduated flask and weigh the bottle again. Dissolve the salt in water, and fill up to the mark. The molarity of the solution can be calculated directly from the weight of salt taken.

2. Indirect method of preparation

If the salt has only been weighed out approximately and the grade potassium dichromate has a purity of less than 99.9%, the solution must be standardized. Take a certain volume of solution $K_2Cr_2O_7$, add excess KI, and H_2SO_4, titrate with $Na_2S_2O_3$ standard solution of known concentration, and determine the endpoint with starch indicator. The reaction formula is as follows:

$$Cr_2O_7^{2-}+6I^-+14H^+ \rightleftharpoons 2Cr^{3+}+3I_2+7H_2O$$

$$I_2+2S_2O_3^{2-} \rightleftharpoons S_4O_6^{2-}+2I^-$$

The concentration of $K_2Cr_2O_7$ the standard solution can be calculated with the following equation:

$$c\left(\frac{1}{6}K_2Cr_2O_7\right)=\frac{(V_1-V_2)\times c(Na_2S_2O_4)}{V} \qquad \textbf{(Equation 7.12)}$$

Where $c\left(\frac{1}{6}K_2Cr_2O_7\right)$ is the concentration of $K_2Cr_2O_7$ standard solution (mol · L^{-1}),

$c(Na_2S_2O_4)$ is the concentration of $Na_2S_2O_4$ standard solution ($mol \cdot L^{-1}$), V_1 is the volume of $Na_2S_2O_4$ standard solution consumed (mL), V_2 is the volume of $Na_2S_2O_4$ standard solution consumed in blank test (mL), and V is the volume of $K_2Cr_2O_7$ standard solution (mL).

7.6.2.3 Application example of potassium dichromate titration

1. Determination of chromium in a chromium (Ⅲ) salt

Chromium (Ⅲ) salts are oxidized to dichromate by boiling with an excess of a persulphate solution in the presence of a little silver nitrate (catalyst). The excess of persulphate remaining after the oxidation is complete is destroyed by boiling the solution for a short time. The dichromate content of the resultant solution is determined by the addition of an excess of a standard iron (Ⅱ) solution and titration of the excess of the latter with standard 0.02 $mol \cdot L^{-1}$ potassium dichromate.

$$2Cr^{3+} + 3S_2O_8^{2-} + 7H_2O \xrightarrow{AgNO_3} Cr_2O_7^{2-} + 6HSO_4^- + 8H^+$$

$$2S_2O_8^{2-} + 2H_2O = O_2\uparrow + 4HSO_4^-$$

Weigh out an amount of salt containing about 0.25 g of chromium and dissolve it in 50 mL of distilled water. Add 20 mL of a 0.1 $mol \cdot L^{-1}$ silver nitrate solution, followed by 50 mL of a 10% ammonium or potassium persulphate solution. Boil the liquid gently for 20 min. Cool and dilute to 150 mL in a graduated flask. Remove 50 mL of the solution with a pipette, and add 50 mL of a 0.1 $mol \cdot L^{-1}$ ammonium iron (Ⅱ) sulfate solution, 200 mL of 1 $mol \cdot L^{-1}$ sulphuric acid, and 0.5 mL of N-phenyl anthranilic acid indicator. Titrate the excess of the iron (Ⅱ) salt with standard 0.02 $mol \cdot L^{-1}$ potassium dichromate until the color changes from green to violet-red.

Standardize the ammonium iron (Ⅱ) sulphate solution against the 0.02 $mol \cdot L^{-1}$ potassium dichromate, using N-phenyl anthranilic acid as an indicator. Calculate the volume of the iron (Ⅱ) solution, which was oxidized by the dichromate originating from the chromium salt, and the percentage of chromium in the sample.

2. Determination of chemical oxygen demand

One significant application of potassium dichromate is in a back-titration for the environmental determination 16 of the amount of oxygen required to oxidize all the organic material in a sample of impure water, such as sewage effluent. This is known as the chemical oxygen demand (COD) and is expressed in terms of milligrams of oxygen required per liter of water, $mg \cdot L^{-1}$. The analysis of the impure water sample is carried out in parallel with a blank determination on pure, double-distilled water.

7.6.3 Iodimetry Titration

7.6.3.1 Overview

The direct iodometric titration method (sometimes termed **iodimetry**) refers to titrations with a standard iodine solution. The indirect iodometric titration method (sometimes termed

iodometry) deals with the titration of iodine liberated in chemical reactions. The normal reduction potential of the reversible system:

$$I_2 + 2e^- \rightleftharpoons 2I^- \qquad \varphi^{\ominus}(I_2/I^-) = 0.536\ V$$

The above equation refers to a saturated aqueous solution in the presence of solid iodine; this half-cell reaction will occur, for example, towards the end of a titration of iodide with an oxidizing agent such as potassium permanganate, when the iodide ion concentration becomes relatively low. Near the beginning, or in most iodometric titrations, when an excess of iodide ion is present, the tri-iodide ion is formed:

$$I_3^- + 2e^- \rightleftharpoons 3I^- \qquad \varphi^{\ominus}(I_3^-/I^-) = 0.536\ V$$

Since iodine is readily soluble in a solution of iodide. The half-cell reaction is better written: $I_3^- + 2e^- \rightleftharpoons 3I^-$, and the standard reduction potential is 0.536 V. Therefore, iodine or the tri-iodide ion is a weaker oxidizing agent than potassium permanganate, potassium dichromate, and cerium (Ⅳ) sulfate.

In most direct titrations with iodine (iodimetry), a solution of iodine in potassium iodide is employed, and the reactive species is the tri-iodide ion, I_3^-. Strictly speaking, all equations involving reactions of iodine should be written with I_3^- rather than with I_2, e. g.

$$I_3^- + 2S_2O_3^{2-} = 3I^- + S_4O_6^{2-}$$

It is more accurate than

$$I_2 + 2S_2O_3^{2-} = 2I^- + S_4O_6^{2-}$$

For the sake of simplicity, however, the equations in this book will usually be written in terms of molecular iodine rather than the tri-iodide ion.

Strong reducing agents (substances with a much lower reduction potential), such as tin (Ⅱ) chloride, sulfurous acid, hydrogen sulfide, and sodium thiosulphate, react entirely and rapidly with iodine, even in an acid solution. With somewhat weaker reducing agents, e. g., arsenic (Ⅲ) or antimony (Ⅲ), the complete reaction occurs only when the solution is kept neutral or very faintly acid; under these conditions, the reduction potential of the reducing agent is a minimum, or its reducing power is maximum.

Suppose a strong oxidizing agent is treated in a neutral or (more usually) acid solution with a large excess of iodide ion. In that case, the latter reacts as a reducing agent, and the oxidant will be quantitatively reduced. In such cases, an equivalent amount of iodine is liberated and titrated with a standard solution of a reducing agent, usually sodium thiosulphate.

The normal reduction potential of the iodine-iodide system is independent of the pH of the solution so long as the latter is less than about 8; at higher values, iodine reacts with hydroxide ions to form iodide and the extremely unstable hypoiodite, the latter being transformed rapidly into iodate and iodide by self-oxidation and reduction:

$$I_2+2OH^- = I^- + IO^- + H_2O$$

$$3IO^- = 2I^- + IO_3^-$$

Two important sources of error in titrations involving iodine are (a) loss of iodine owing to its appreciable volatility and (b) acid solutions of iodide are oxidized by oxygen from the air:

$$4I^- + O_2 + 4H^+ = 2I_2 + 2H_2O$$

In the presence of excess iodide, the volatility is decreased markedly through the formation of the tri-iodide ion; at room temperature, the loss of iodine by volatilization from a solution containing at least 4% of potassium iodide is negligible, provided the titration is not prolonged unduly. Titrations should be performed in cold solutions in conical flasks, not open beakers. If a solution is to stand, it should be kept in a glass-stoppered vessel. The atmospheric oxidation of iodide is negligible in neutral solution without catalysts, but the oxidation rate increases rapidly with decreasing pH. The reaction is catalyzed by certain metal ions of variable charge value (particularly copper), nitrite ions, and strong light.

For this reason, titrations should not be performed in direct sunlight, and solutions containing iodide should be stored in amber glass bottles. Furthermore, the air oxidation of the iodide ion may be induced by the reaction between the iodide and the oxidizing agent, especially when the main reaction is slow. Solutions containing an excess of iodide and acid must, therefore, not be allowed to stand longer than necessary before titration of the iodine. Suppose prolonged standing is necessary (as in the titration of vanadate or Fe^{3+} ions). In that case, the solution should be free from air before the addition of iodide and the air displaced from the titration vessel by carbon dioxide [e. g., by adding small portions (0. 2-0. 5 g) of pure sodium hydrogen carbonate to the acid solution, or a little solid carbon dioxide, dry ice]; potassium iodide is then introduced and the glass stopper replaced immediately.

It seems appropriate to refer to the uses of a standard solution containing potassium iodide and potassium iodate. This solution is quite stable and yields iodine when treated with acid.

$$IO_3^- + 5I^- + 6H^+ = 3I_2 + 3H_2O$$

7. 6. 3. 2 Preparation of the standard solution

The standard solution is prepared by dissolving a weighed amount of pure potassium iodate in a solution containing a slight excess of pure potassium iodide and diluting it to a definite volume. This solution has two important uses. The first is as a source of a known quantity of iodine in titrations; it must be added to a solution containing strong acid; it cannot be employed in a neutral medium or low acidity.

The second use is in determining the acid content of solutions iodometrically or in the standardization of solutions of strong acids. It is evident from the above equation that the amount of iodine liberated is equivalent to the acid content of the solution. Thus, if, 25 mL of an approximately 0.1 mol • L^{-1} solution of a strong acid is treated with a slight excess of potassium iodate (25 mL of 0.02 mol • L^{-1} potassium iodate solution) and a slight excess of potassium iodide solution (10 mL of a 10% solution), and the liberated iodine titrated with standard 0.1 mol • L^{-1} sodium thiosulphate with the aid of starch as an indicator, the acid concentration may be readily evaluated.

7.6.3.3 Detection of the endpoint

A solution of iodine in aqueous iodide has an intense yellow-to-brown color. One drop of 0.05 mol • L^{-1} iodine solution imparts a perceptible pale yellow color to 100 mL of water so that iodine can serve as its indicator in otherwise colorless solutions. The test is made much more sensitive by using a starch solution as an indicator. Starch reacts with iodine in the presence of iodide to form an intensely blue-colored complex, visible at deficient concentrations of iodine. The sensitivity of the color reaction is such that a blue color is visible when the iodine concentration is 2×10^{-5} mol • L^{-1} and the iodide concentration is greater than 4×10^{-4} mol • L^{-1} at 20 ℃. The color sensitivity decreases with increasing temperature of the solution; thus, at 50 ℃, it is about ten times less sensitive than at 25 ℃. The sensitivity decreases upon adding solvents, such as ethanol: no color is obtained in solutions containing 50% ethanol or more. It cannot be used in a strongly acid medium because hydrolysis of the starch occurs.

Starches can be separated into two major components, amylose and amylopectin, which exist in different proportions in various plants. Amylose, a straight-chain compound abundant in potato starch, gives a blue color with iodine, and the chain assumes a spiral form. Amylopectin, which has a branched-chain structure, forms a red-purple product, probably by adsorption.

The great merit of starch is that it is inexpensive. It possesses the following disadvantages: ① insolubility in cold water; ② instability of suspensions in water; ③ it gives a water-insoluble complex with iodine, the formation of which precludes the addition of the indicator early in the titration (for this reason, in titrations of iodine, the starch solution should not be added until just before the endpoint when the color begins to fade); and ④ there is sometimes a "drift" endpoint, which is marked when the solutions are dilute.

7.6.3.4 Application example of iodimetry titration

1. Determination of copper in crystallized copper sulfate

Weigh about 3.0 g of the salt, dissolve it in water, and fill up to 250 mL in a graduated flask. Shake well. Pipette 50.0 mL of this solution into a 250 mL conical flask, add 1 g potassium iodide (or 10 mL of a 10% solution) (note 1), and titrate the liberated iodine with standard 0.1 mol • L^{-1} sodium thiosulphate (note 2). Repeat the titration with two other 50 mL portions of the copper sulfate solution.

The reaction, written in molecular form, is:

$$2CuSO_4 + 4KI = 2CuI + I_2 + 2K_2SO_4$$

From which it follows that:

$$2CuSO_4 \equiv I_2 \equiv 2Na_2S_2O_3$$

Notes:

(1) If free mineral acid is present in a similar determination, a few drops of dilute sodium carbonate solution must be added until a faint permanent precipitate remains, which is removed by a drop or two of acetic acid. The potassium iodide is then added, and the titration continued. The solution should have a pH of 4.0-5.5 for accurate results.

(2) After the addition of the potassium iodide solution, run in standard 0.1 $mol \cdot L^{-1}$ sodium thiosulphate until the brown color of the iodine fades, then add 2 mL of starch solution, and continue adding the thiosulphate solution until the blue color commences to fade. Then add about 1 g of potassium thiocyanate or ammonium thiocyanate, preferably as a 10% aqueous solution; the blue color will instantly become more intense. Complete the titration as quickly as possible. The precipitate is pale pink, and a distinct permanent endpoint is readily obtained.

2. Analysis of hydrogen peroxide

Hydrogen peroxide reacts with iodide in an acid solution following the equation:

$$H_2O_2 + 2H^+ + 2I^- = I_2 + 2H_2O$$

The reaction velocity is comparatively slow but increases with increasing acid concentration. Adding three drops of a neutral 20% ammonium molybdate solution renders the reaction almost instantaneous. Still, as it also accelerates the atmospheric oxidation of the hydriodic acid, the titration is best conducted in an inert atmosphere (nitrogen or carbon dioxide).

The iodometric method has the advantage over the permanganate method in that it is less affected by stabilizers, which are sometimes added to commercial hydrogen peroxide solutions. These preservatives are often boric acid, salicylic acid, and glycerol and render the results obtained by the permanganate procedure less accurate.

7.7 Calculation Practice of Redox Titration

The chemical reactions of redox titration are rather complicated, so in the relevant calculation, first of all, we must know the stoichiometry between the titrator and the substance titrated, select the basic unit of titrator, determine the basic unit of the component to be measured, and then calculate it according to the mol of substance rules.

Example 7. 8

The sample of 5. 860 0 g NaClO weighed precisely was put into a volumetric bottle, and the volume was set to 250 mL. From it, 25. 00 mL solution was taken out and transferred into an iodine flask, then coving tightly for a moment. The 25. 00 mL NaClO solution was titrated with a standard solution $Na_2S_2O_3$ to the endpoint using starch as an indicator and 20. 64 mL $Na_2S_2O_3$ the solution was used. Calculate the mass fraction of Cl in the sample. Known about $T(I_2/Na_2S_2O_3)=0.013\,35\ g \cdot mol^{-1}$, $M(I_2)=253.8\ g \cdot mol^{-1}$, $M(Cl)=35.45\ g \cdot mol^{-1}$.

Solution:

According to the question context, the reaction equation can be written as follows:

$$2ClO^- + 4H^+ = Cl_2 + 2H_2O$$

$$Cl_2 + 2I^- = 2Cl^- + I_2$$

$$I_2 + 2S_2O_3^{2-} = S_4O_6^{2-} + 2I^-$$

According to the above reaction equations, confirming the basic unit $\frac{1}{2}I_2$ for I_2 and Cl for Cl.

$$c(Na_2S_2O_3) = \frac{T(I_2/Na_2S_2O_3) \times 10^3}{M_{\frac{1}{2}I_2}}$$

$$c(Na_2S_2O_3) = \frac{0.013\,35 \times 10^3}{126.9} = 0.105\,2(mol \cdot L^{-1})$$

$$\omega(Cl) = \frac{c(Na_2S_2O_3) \times V(Na_2S_2O_3) \times M(Cl)}{m(NaClO) \times \frac{25.00}{250.00} \times 1\,000}$$

$$= \frac{0.105\,2 \times 20.64 \times 35.45}{5.860\,0 \times \frac{25.00}{250.00} \times 1\,000} = 0.131\,4$$

The results show that the mass fraction of Cl in the sample is 0. 131 4.

Example 7. 9

The sample of 0. 387 8 g $Na_2SO_3 \cdot 5H_2O$ was weighed precisely and dissolved into water, and 50. 00 mL of the solution of I_2 was added to the sample to react with it. The standard solution $Na_2S_2O_3$ was used to titrate remained mixture matters until the titration endpoint when 25. 40 mL solution of $Na_2S_2O_3$ was used. Calculate the mass fraction of Na_2SO_3. Known about $c\left(\frac{1}{2}I_2\right)=0.097\,70\ mol \cdot L^{-1}$, $c(Na_2S_2O_3)=0.100\,8\ mol \cdot L^{-1}$, $M(Na_2S_2O_3)=126.05\ g \cdot mol^{-1}$.

Solution:

According to the question context, the reaction equation can be written as follows:

$$I_2 + SO_3^{2-} + H_2O = 2H^+ + 2I^- + SO_4^{2-}$$

$$S_2O_3^{2-} + I_2 = S_4O_6^{2-} + 2I^-$$

According to the above reaction equations, confirming the basic unit $\frac{1}{2}Na_2SO_3$ for Na_2SO_3.

$$\omega(Na_2SO_3) = \frac{\left[c\left(\frac{1}{2}I_2\right) \times V(I_2) - c(Na_2S_2O_3) \times V(Na_2S_2O_3)\right] \times M\left(\frac{1}{2}Na_2SO_3\right)}{m(Na_2SO_3) \times 1\ 000}$$

$$= \frac{(0.097\ 70 \times 50.00 - 0.100\ 8 \times 25.40) \times 63.02}{0.387\ 8 \times 1\ 000} = 0.377\ 8$$

The results show that the mass fraction of Na_2SO_3 in the sample is 0.377 8.

Key Glossary

Oxidation-reduction (redox) titrations are based on an oxidation-reduction reaction between analytic and titrant and are similar in setup to acid-base titrations, but the focus is different. In contrast, acid-base titrations follow the movement of protons. Redox titrations follow the transfer of charge (as electrons) to reach the equivalence point.

Standard electrode potential is the measure of the individual potential of a reversible electrode at standard state, which is with solutes at an effective concentration of $a_{Ox} = a_{Red} = 1\ mol \cdot L^{-1}$, At 298.15 K, and gases at a pressure of 101.325 kPa.

Condition electrode potential means the electrode potential under special conditions of $c_{Ox} = c_{Red} = 1\ mol \cdot L^{-1}$.

Internal oxidation-reduction indicators are compounds that exhibit different colors in oxidized and reduced forms.

Self-indicating reagents are well illustrated by potassium permanganate, one drop of which will impart a visible pink coloration to several hundred milliliters of solution, even in the presence of slightly colored ions.

The specific indicator is itself not redox but can react with an oxidant or reductant to produce a color reaction. Moreover, the color reaction is irreversible, so that the titration endpoint can be indicated.

Essential Exercises

Review Questions

1. What is the essence of redox reaction? What are the characteristics of redox reactions?

2. What are the factors affecting the rate of redox reaction?

3. What are the commonly used indicators for redox titration?

4. What are the commonly used titrants in redox titration?

5. Why is Iodine almost always used in a solution containing excess I^-?

6. State one way to make a standard triiodide solution.

7. In which technique, iodimetry or iodometry, is the starch indicator not added until just before the endpoint? Why?

Cumulative Problems

1. A 20.0 mL solution of 0.005 00 mol • L^{-1} Sn^{2+} in 1 mol • L^{-1} HCl was titrated with 0.020 0 mol • L^{-1} Ce^{4+} to give Sn^{4+} and Ce^{3+}. Calculate the potential at the following volumes of Ce^{4+}: 0.100 mL, 1.00 mL, 5.00 mL, 9.50 mL, 10.00 mL, 10.10 mL, and 12.00 mL. Sketch the titration curve.

2. A solid mixture weighing 0.054 85 g contained only ferrous ammonium sulfate [$FeSO_4(NH_4)_2SO_4 \cdot 6H_2O$; $M=392.13$] and ferrous chloride ($FeCl_2 \cdot 6H_2O$; $M=234.84$). The sample was dissolved in 1 mol • L^{-1} H_2SO_4 and the Fe^{2+} required 13.39 mL of 0.012 34 mol • L^{-1} Ce^{4+} for complete oxidation to Fe^{3+}. Calculate the weight percent of Cl in the original sample.

3. Consider the titration of 100.0 mL of 0.010 0 mol • L^{-1} Ce^{4+} in 1 mol • L^{-1} $HClO_4$ by 0.040 0 mol • L^{-1} Cu^+ to give Ce^{3+} and Cu^{2+}.

①Write a balanced titration reaction.

②Write two different half-reactions for the indicator electrode.

③Write two different Nernst equations for the cell voltage.

④Calculate φ at the following volumes of Cu^+: 1.00 mL, 12.5 mL, 24.5 mL, 25.0 mL, 25.5 mL, 30.0 mL, and 50.0 mL.

4. Consider the titration of 25.0 mL of 0.010 0 mol • L^{-1} Sn^{2+} with 0.050 0 mol • L^{-1} Tl^{3+} in 1 mol • L^{-1} HCl.

①Write a balanced titration reaction.

②Write two different half-reactions for the indicator electrode.

③Write two different Nernst equations for the cell voltage.

④Calculate φ at the following volumes of Tl^{3+}: 1.00 mL, 2.50 mL, 4.90 mL, 5.00 mL, 5.10 mL, and 10.0 mL.

5. Consider the titration of 25.0 mL of 0.050 0 mol • L^{-1} Sn^{2+} with 0.100 mol • L^{-1} Fe^{3+} in 1 mol • L^{-1} HCl to give Fe^{2+} and Sn^{4+}.

①Write a balanced titration reaction.

②Write two half-reactions for the indicator electrode.

③Write two Nernst equations for the cell voltage.

④Calculate φ at the following volumes of Fe^{3+}: 1.0 mL, 12.5 mL, 24.0 mL, 25.0 mL, 26.0 mL, and 30.0 mL.

6. When 25.00 mL of unknown were passed through a Jones reductant, molybdate ion (MoO_4^{2-}) was converted into Mo^{3+}. The filtrate required 16.43 mL of 0.010 33 mol · L^{-1} $KMnO_4$ to reach the purple endpoint.

$$MnO_4^- + Mo^{3+} \longrightarrow Mn^{2+} + MoO_2^{2+}$$

A blank required 0.04 mL. Balance the reaction and find the molarity of molybdate in the unknown.

7. Preparing 1 L sodium thiosulfate solution $c(Na_2S_2O_3)=0.200\ 0$ mol · L^{-1}. How many grams of sodium thiosulfate will be taken?

8. Weighing 0.193 6 g pure potassium dichromate solid precisely, which was dissolved in water and acidified, and then, excessive potassium iodide was added into the solution. It was titrated with sodium thiosulfate solution to the end of titration when the volume of sodium thiosulfate solution was assumed at 33.61 mL. Calculate the molarity of $Na_2S_2O_3$.

9. 0.186 1 g $KHC_2O_4 \cdot H_2O$ can react exactly with 30.00 mL $KMnO_4$ solution in an acidic environment. Calculate the molarity of $KMnO_4$.

Chapter 8

Absorption Spectrophotometry

Key Points

- *Principle of Absorption Spectrophotometry*
- *Mathematical Expression and the Significance of Beer-Lambert Law*
- *Role of Absorption Curve in Absorption Spectrophotometry*
- *Plotting Method of the Standard Curve*
- *Common Spectrophotometer and Operation Method*
- *Color Reaction*

8.1 Outline

In chemistry, **absorption spectrophotometry** is the quantitative measurement of the selective absorption properties of the test material as a function of wavelength. It includes colorimetry, visible spectrophotometry, UV spectrophotometry, and infrared spectroscopy. This chapter mainly focuses on absorption spectrophotometry with visible light.

Most substances show different colors. For instance, the potassium permanganate solution is purple-red. The different shades of a solution are related to the concentration of colored materials. The higher the concentration, the darker the color is. Thus, by comparing the shade of the solution, the concentration of the colored substance in the solution could be detected, and this method is named colorimetry. Almost all the elements could be analyzed in colorimetry. With the development of test instruments, the spectrophotometer has been widely used in colorimetry since the 1940s. The analytical method using a spectrophotometer is called **spectrophotometry**.

8.1.1 Characteristic and Application of Absorption Spectrophotometry

Compared with titrimetric analysis, absorption spectrophotometry mainly has the following characteristics.

(1) High sensitivity Absorption spectrophotometry has high sensitivity. It could be used in microcomponents measurement, with a low concentration limit of 10^{-6}-10^{-5} $mol \cdot L^{-1}$. If the sample is enriched before detection, the sensitivity could be improved to one or two orders of magnitude.

(2) High accuracy The relative error of macroanalysis is permille, while the microanalysis normally is a percentage. The relative error of colorimetry is 5%-10%, and spectrophotometry is 2%-5%, which can meet the requirements of microanalysis. If the precise spectrophotometer is used in measuring, the relative error could be decreased to 1%-2%, and even to 0.2%-0.5%.

(3) Quick and simple The instrument is simple and easy to operate. Since highly sensitive and selective color agents and screening agents have constantly appeared in recent years, colorimetry or spectrophotometry could be used directly without separation, which shortens the analysis time and quickens the test speed.

(4) Wide application Colorimetry or spectrophotometry could measure most inorganic ions and organic compounds directly or indirectly. At present, spectrophotometry is commonly used in agricultural routine analysis.

8.1.2 Basic Properties of Light

Light is a form of electromagnetic radiation. Thus, it shows the features of electromagnetic waves, including wave and particle properties. The wave property of light is confirmed by interference, diffraction, and refraction phenomena. It is usually described in terms of wavelength, frequency, and velocity, and the relation is shown as follows:

$$V=\frac{c}{n}=\lambda\times\nu$$

where c is the velocity of light in a vacuum, which is a physical constant in many fields of physics, and its value is a constant around $2.997\ 924\ 58\times10^{8}$-$3.0\times10^{8}\ \mathrm{m\cdot s^{-1}}$; λ is the wavelength, and its units are usually in meter (m), centimeter (cm), millimeter (mm), and nanometer (nm); ν is the frequency and its unit is Hz; n is the refractivity (in a vacuum it is 1); V is the velocity of light.

Electromagnetic waves have a wide range of wavelengths. As shown in Table 8.1, the wavelength of X-rays is shorter, while the wavelength of radio waves is longer. Visible light accounts for only a small part of electromagnetic waves.

Table 8.1 Electromagnetic spectrum

Region of the spectrum	Wavelength range
X-rays	0.1-10 nm
Far ultraviolet	10-200 nm
Near ultraviolet	200-400 nm
Visible	400-750 nm
Near-infrared	0.75-2.5 μm
Mid-infrared	2.5-5.0 μm

Table 8.1 continued

Region of the spectrum	Wavelength range
Far infrared	5.0-1 000 μm
Micro waves	0.1-100 cm
Radio waves	1-1 000 m

The photoelectric effect, light pressure phenomenon, and photochemical effect could confirm the particle property of light, which could take light as the composition of particles (photons or quantum photons) with a certain energy. The energy depends on the frequency or wavelength of light, and the relation is:

$$E = h \times \nu$$

where h is Planck's constant (6.626×10^{-34} J · S); ν is the frequency; E is the energy of photons (unit with J or eV).

A certain wavelength of light possesses a certain amount of energy, and the longer the wavelength (the lower frequency), the lower the energy of photons.

8.1.3 Selective Absorption of Light and the Color of Solution

White light is a combination with a certain proportion of seven-colored lights, including red, orange, yellow, green, cyan, blue, and purple, in which the wavelength range of all these colored lights is different proportions (Table 8.2). Thus, the wavelength range of white light is almost the entire visible spectrum of 400-760 nm. The light composed of different wavelengths is called **polychromatic light**, and the light of a single wavelength is called **monochromatic light.** Suppose two colored monochromatic lights could be mixed proportionately to form white light. In that case, the pair of colored monochromatic lights are called **complementary color light**, which is complementary to light, as shown in Figure 8.1.

Table 8.2 Approximate wavelengths of colors

Color	Wavelength range/nm
Ultraviolet	< 400
Violet	400-450
Blue	450-470
Cyan	470-500
Green	500-570
Yellow	570-590
Orange	590-620
Red	620-760
Infrared	> 760

The different colors presented in the solution are caused by the selective absorption of the colored substance to the light with a certain wavelength. The rays with rest wavelengths are transmitted. Above all, the solution exhibits the color of transmission lights complemented by absorbed lights. If an opaque substance presents white, all light wavelengths are reflected equally. If an opaque substance presents black, all wavelengths of light are transmitted equally. For example, the copper sulfate solution presents a blue color because it absorbs the yellow light in white light, and the potassium permanganate solution presents a purple color due to absorbing the green light in white light.

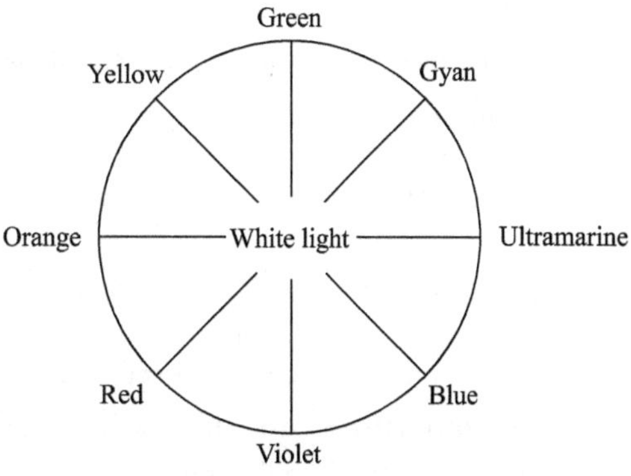

Figure 8.1 The complementation diagram of the light

8.1.4 Absorption Curve

To accurately express the absorption degree of the solution with different wavelengths of light, generally taking the absorption degree (absorbance) as the Y-axis and the wavelength as the X-axis, the **absorption spectra curve** (short for **absorption curve**) is established. The absorption curve should have a maximum absorption peak in the solution of each substance. The wavelength with the corresponding peak is called the maximum absorption wavelength, which is denoted in λ_{max}. For example, as shown in Figure 8.2, the potassium permanganate solution has the maximum absorption wavelength for green light with a wavelength of 525 nm. In contrast, it has little absorption of purple and red light. Therefore, the potassium permanganate solution presents a purple-red color. As shown in Figure 8.2, although the height of the absorption curve is different for potassium permanganate solution with different concentrations, the λ_{max} is constant.

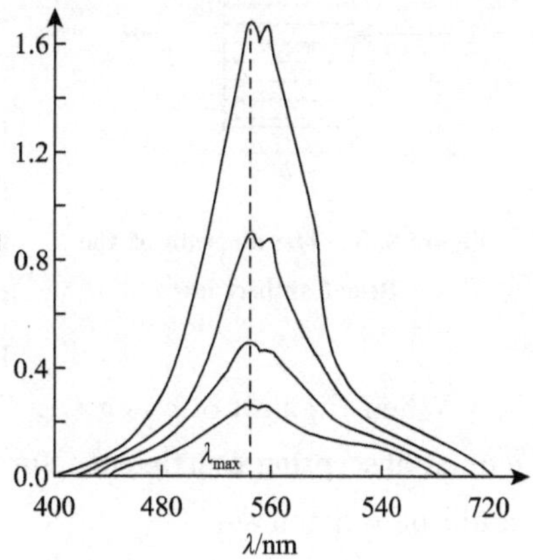

Figure 8.2 The absorption curve of $KMnO_4$ solution

8.2 Beer-Lambert Law

When a parallel beam of monochromatic light passes through a sample of thickness b and concentration c, the incident irradiance I_o could decrease to the transmitted irradiance I_t ($I_t \leqslant I_o$). Some light radiation might be absorbed by the sample (Figure 8.3). Transmittance (T) is expressed as the ratio of transmitted light intensity to the incident light intensity. To accurately state the degree of the radiation attenuation of the light, absorbance had been put forward. **Absorbance (*A*)** is the negative of the transmittance in logarithmic.

Transmittance: $$T=\frac{I_t}{I_o}$$ **(Equation 8.1)**

Absorbance: $$A=-\lg T=-\lg\frac{I_t}{I_o}=\lg\frac{I_o}{I_t}$$ **(Equation 8.2)**

In 1760, Lambert discovered that the degree of the radiation attenuation of the light is directly proportional to the path length of the medium when light passes through a transparent medium. In 1852, Beer stated that the absorption extent is proportional to the concentration of the absorbing substance in the solution.

According to the finding above, the absorbance of a solution is directly proportional to its concentration and the path length of the absorbing solution when a parallel beam of monochromatic light vertically irradiates the sample solution-called **Beer-Lambert law.** The corresponding mathematical equations are shown as follows:

$$A=\lg\frac{I_o}{I_t}=kbc$$ **(Equation 8.3)**

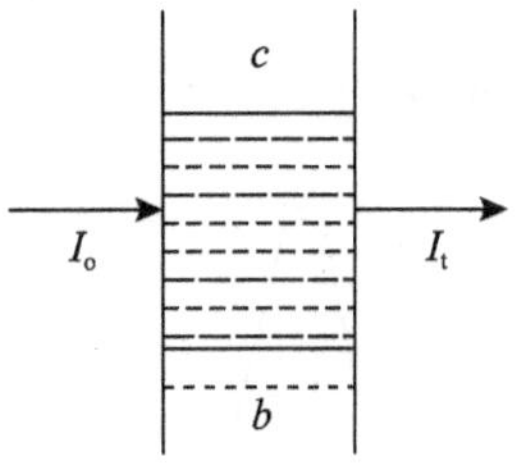

Figure 8.3 The diagram of the Beer-Lambert law

In the above equation, *b* is the pathlength of the absorbing solution expressed in centimeters (Figure 8.3); *c* is the concentration expressed in grams per liter or moles per liter; *k* is the proportionality constant, which is the absorbance of the absorbing substance in unit concentration and unit pathlength. Under a certain condition (monochromatic light, solvent, and temperature are constant), *k* is a constant value depending on the unit of concentration and path length.

When the unit of *c* is $g \cdot L^{-1}$, and *b* is cm, the proportional constant *k* is expressed in *a* named **absorption coefficient**, and its unit is $L \cdot g^{-1} \cdot cm^{-1}$. The above absorbance equation could be written as:

$$A=abc$$ **(Equation 8.4)**

When the unit of *c* is $mol \cdot L^{-1}$ and *b* is cm, the proportional constant *k* is expressed in ε, named the **molar absorption coefficient**, and its unit is $L \cdot mol^{-1} \cdot cm^{-1}$. The above absorbance equation could be written as:

$$A=\varepsilon bc$$ **(Equation 8.5)**

ε is the characteristic constant of absorption material in a certain wavelength and medium, reflecting the absorption material's sensitivity. The higher the ε, the greater the absorbance.

Example 8.1

The ion (Ⅱ) solution concentration is 1.0×10^{-3} g · L^{-1}, and it could react with phenanthroline to form an orange-red complex. However, the absorbance of the complex is 0.20 when the wavelength is 508 nm, and the cuvette thickness is 1.0 cm. Therefore, please calculate the molar absorbance constant of phenanthroline.

Solution:

$$M(\mathrm{Fe})=55.85$$

$$c(\mathrm{Fe}^{2+})=\frac{1.0\times10^{-3}}{55.85}=1.8\times10^{-5}(\mathrm{mol}\cdot\mathrm{L}^{-1})$$

$$\varepsilon=\frac{A}{bc}=\frac{0.20}{1.0\times1.8\times10^{-5}}=1.1\times10^{4}(\mathrm{L}\cdot\mathrm{mol}^{-1}\cdot\mathrm{cm}^{-1})$$

8.3 Application of Beer-Lambert Law

Beer-Lambert law could be used in several different measurements: ① The molar absorption coefficient of a substance could be calculated if the absorbance, concentration, and path length are known. ② The absorbance value could also be calculated if the concentration, absorption coefficient, and path length are known. ③In quantitative analysis by spectrometry, the assay measurement of the unknown solution concentration, including the comparative and standard curve methods (in Section 8.4.2), all depend on the Beer-Lambert law to get the absorbance or standard curve.

Beer-Lambert law can apply to visible, infrared, and UV light. It is also not just for the solution with isotropic-non-scattering but for solids and gas. Therefore, it is the basis of absorption spectrophotometry and has wide applications.

8.4 Color Reaction

8.4.1 Color Reaction and Color Reagent

In chemistry, a **color reaction** is a chemical reaction that transforms colorless chemical compounds into colored derivatives, which can be detected visually or with the aid of a colorimeter. A colorimeter cannot usually determine the concentration of a colorless solution. The addition of a color reagent leads to a color reaction, and the absorbance of the colored product can then be measured with a colorimeter. If the color of the test substance is darker, it could be measured directly. If the color of the test substance is colorless or light-colored, the color reagent is needed to react with test ions to produce a colored compound. The key to color reaction is that the color should be specific for the test component.

1. Requirements of the color reaction

(1) High sensitivity The darker color of the colored compound generated could result in higher reaction sensitivity. Thus, the colored substance with a high molar absorption constant must be selected, and the light with the corresponding maximum absorption wavelength could be taken as the incident light. The molar absorption constant generally means the absorption constant at the maximum absorption wavelength. From the molar absorption constant of colored substances generated by the reaction between the different color reagents with the same ion, we could compare the sensitivity of the color reagents for the test ions. As shown in Table 8.3, the ε of Cu^{2+} reacted with PAN is the maximum value compared with others, and this reaction is more sensitive. Therefore, the chromogenic reagent PAN should be used to determine the micro Cu^{2+}.

Table 8.3 The reaction of the ε of Cu^{2+} with different color reagents

Color reagents	λ_{max}/nm	ε_{max}/$(L \cdot mol^{-1} \cdot cm^{-1})$
Ammonia	620	1.2×10^2
Sodium diethyldithiocarbamatre (DDTC)	436	1.3×10^4
1-(2-pyridinylazo)-2-Naohthalenol (PAN)	550	4.5×10^4

(2) High selectivity Choosing a color reagent that only reacts with the test component is better. If other components also could react with the color reagent to form the colored compound, the masking agent or other methods should be adopted to exclude the interference. If necessary, the interference component should be separated in advance.

(3) The fixed component of the colored compound The colored compound generated must have a fixed component. Thus, the test substance has a quantification relation with the colored compound.

(4) The high stability of the colored compound The more stable the colored compound, the less interference from other ions. The colored compound couldn't dissociate and associate in test time.

2. Type of color reaction

The reaction between color reagents and test components can be classified into complexation reaction, oxidation-reduction reaction, ion-association reaction, salification reaction, discoloring reaction, adsorption color reaction, etc. The complexation reaction is one of the most widely used reactions.

3. Color reagent

Various color reagents have been developed to determine the concentrations of different substances. However, there are a few inorganic color reagents due to the generated complex being unstable, insensitive, and unselected. For instance, iron, molybdenum, tungsten, and niobium could be measured by color reaction with KSCN; silicon, phosphorus, and vanadium are measured by ammonium molybdate, while H_2O_2 measures titanium.

The organic color reagent contains chromophores and auxochromes. The chromophore is the group with unsaturated bonds, such as azo groups, p-quinonyl, carbonyl, etc. π electron needs less energy when excited, and it can absorb the visible light above 200 nm wavelength for coloration. The auxochrome is the group with the lone pair electron, as amidogen, hydroxy, halogen-groups, etc. These groups reacted with the unsaturated bonding of chromophores, which could make the color darker.

8.4.2 Influencing Factors of Color Reaction

Influencing factors of a color reaction include the solution acidity, dose of color agent, the adding order of reagents, color time, color temperature, the stability of organic complexes, the interference of coexistent ions, etc.

1. The solution acidity

The color reaction is mostly a complex reaction in which the acid effect is the main side reaction. As the acid effect is related to acidity, the solution's acidity could be one factor of color reaction. Influences of the acidity change in the color reaction mainly include the equilibrium concentration and the color of the color reagent, the existential state of the test metal ion, and the composition of the complex. In addition, the relation curve of pH can determine the appropriate pH range of color reactions with absorbance.

The acidity greatly influences the color reaction; thus, the appropriate acidity of a certain color reaction must be determined by experiments. For example, several test ion solutions with the same concentration could be colored in buffer solutions with different pH, and then the absorbance should be measured. The curve is established by taking the pH as the X-axis and the absorbance as the Y-axis (Figure 8.4). The corresponding pH range of the straight part of the curve is the most appropriate acidity range.

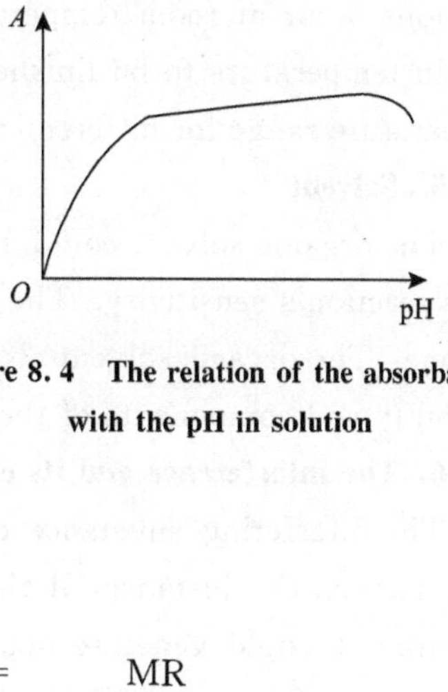

Figure 8.4 The relation of the absorbance with the pH in solution

2. The dosage of the color reagent

$$\underset{\text{(test component)}}{\mathrm{M}} + \underset{\text{(color reagent)}}{\mathrm{R}} = \underset{\text{(colored complex)}}{\mathrm{MR}}$$

The excess color reagent should be added to the solution to complete the color reaction fully. However, this does not mean that the more color reagents, the better the color reaction. The more color reagent might cause an adverse reaction for certain colors, which is bad for the measurement. The result of experiments in practical work should determine the dosage of color reagents. The color reagent with different dosages is added to a series of sample solutions with the same concentration, and then the absorbance of the solution is measured respectively. If the series dosage of color reagents measures the absorbance within a certain range and keeps a constant value, as shown in Figure 8.5A, the dosage of the color reagent could be determined in this range.

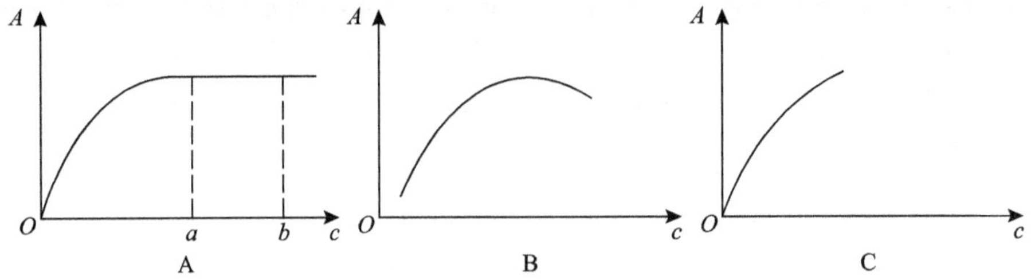

Figure 8.5 The relation of the test solution absorbance with the color reagent concentration

3. Color reaction time

Some color reactions occur almost instantaneously. The solution color quickly reaches a stable state and stays stable for a long time. Although some reactions finish quickly, they may have complex colors, resulting in color fading. Some color reactions occur slowly, and the solution color is steady over time. The curve of absorbance time should determine the appropriate time.

4. Color reaction temperature

The different color reactions have different temperature requirements. Most of the color reactions occur at room temperature. However, some color reactions must be heated to a certain temperature to be finished. Thus, the experiment should determine the appropriate temperature range for different reactions.

5. Solvent

The organic solvent could decrease the organic compound's dissociation and increase the color reaction's sensitivity. The color of the $Fe(SCN)_3$ solution could be deeper by adding acetone. The organic solvent also might improve the rate of color reactions and influence the solubility and components of the colored complex.

6. The interference and its elimination methods

The interfering substance contained in the solution could influence the measurement components. For instance, if the interfering substance has color itself or reacts with color reagents, it could generate absorption in measured conditions, which results in positive interference. The interfering substance reacted with the test component, which caused the uncompleted reaction and could result in interference. The absorbance of the solution could not be exactly determined because the interfering substance extracted from the solution in the measurement condition makes the solution turbid. Certain interference could be eliminated by the methods as follows in experiments:

①Control of the solution acidity.

②Adding the masking agent. The selected condition contains that the masking agent does not react with the test ion, and the color of the complex formed by the masking agent with the interference could not interfere with the measurement of ions.

③Changing the valence state of the interference ion by oxidation-reduction reactions.

④Use the correction coefficient.

⑤The reference solution could eliminate the interference of the color reagent and certain coexisting colored ions.

⑥Selecting the appropriate wavelength.

⑦The dosage of the color reagent could be added to eliminate interferences when there are interference ions that could expand the color reagent in the solution.

⑧Separation.

The pro-separation method, including solvent extraction, precipitation, electrolysis, ionic exchange, etc., should be adopted when the above techniques have no effects. The test ion should be separated from various interference ions before being measured.

8.5 Absorbance Spectrophotometry and the Instrument

Depending on the instrument used, absorbance spectrophotometry could fall into three categories: visual colorimetry, photoelectric colorimetry, and spectrophotometry.

8.5.1 Visual Colorimetry

Visual colorimetry is an analytical method that could determine the material content by comparing the solution's color depth with the eyes. The common visual colorimetry is the standard series method. The experiment method is as follows: **Nessler tube** is usually used, a colorless glass tube with a uniform cross-section and a flat bottom. The standard solution with a series of different quantities and the test solution must be added to a set of Nessler tubes, respectively. An equal number of colors and the other reagents are added into the Nessler tubes under the same experiment condition. The solution in tubes should be diluted to a certain volume. Then, the color intensity of test solutions and standard solutions could be compared using vertical observation tubes. If the test solution has the same color as one standard solution, both concentrations are the same. If the color of the test solution is between two successive standards, the average arithmetic value of the two standards could be taken as the concentration of the test solution.

The shortcoming of visual colorimetry is mainly the low accuracy. The concentration could not be measured if the other colored material existed in the test solution. Moreover, since the colors of various colored solutions are unstable, standard series solutions that can't be stored for a long time need to be prepared for determination. Although the permanent standard series solution could be prepared by some certain colored materials (such as potassium dichromate, copper sulfate, cobaltous sulfate, etc.), or the permanent level could be prepared by the colored plastic and the colored glass, they need correction caused by the color difference with the test solution.

The advantages of visual colorimetry include the simplest apparatus, easy operation, application to analyze many samples, and high sensitivity. The experiment condition is polychromatic light-white light, so visual colorimetry could also measure some color reactions that do not conform to Beer-Lambert law. Thus, visual colorimetry is widely used in routine analysis with undemanding accuracies, such as rapid nitrogen, phosphorus, and potassium determination in soil and plants.

8. 5. 2 Spectrophotometry

1. Principle

Spectrophotometry can measure the absorbance of the solution with a spectrophotometer and determine the concentration of the test solution according to Beer-Lambert law. However, spectrophotometry and visual colorimetry are not the same in principle. Spectrophotometry is used to compare the absorbance of the colored solutions for a certain wavelength of light, while visual colorimetry is used to compare the intensity of transmitted light. For example, the concentration of $KMnO_4$ in the solution is determined by two methods. The absorbance of $KMnO_4$ solution for yellow-green light is determined by spectrophotometry. On the other hand, the intensity of $KMnO_4$ solution transmitting red-purple light is determined by visual colorimetry.

2. Assay method

(1) Comparative method For the test solution of the absorbing substance with concentration c_x, consider the standard solution of the same substance with concentration c_s. It's better to guarantee the c_s is close to the c_x. Then, these two concentration solutions are placed in a spectrophotometer with the same measurement condition. The corresponding absorbance A_s and A_x are detected by the color reaction. Depend on Beer-Lambert law are as follows:

$$A_s = \varepsilon b_s c_s; \ A_x = \varepsilon b_x c_x$$

The phase ratio of the above formulas is: $\dfrac{A_s}{A_x} = \dfrac{\varepsilon b_s c_s}{\varepsilon b_x c_x}$

And finally: $c_x = \dfrac{A_s}{A_x} c_s$

It should be noted that the result is reliable only when the measured value c_x is close to the standard value c_s in the calculation. Otherwise, there may be a significant error.

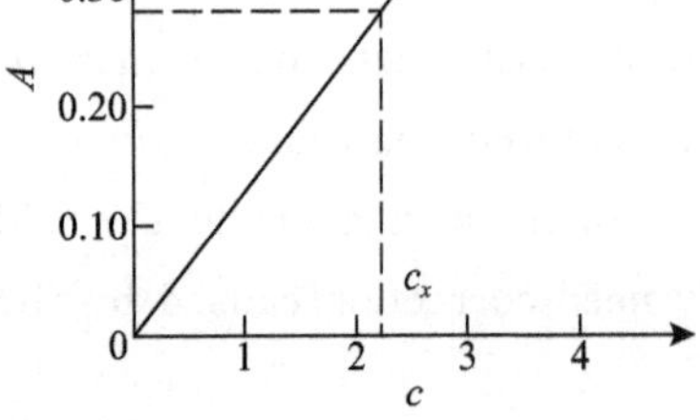

Figure 8. 6 Standard curve

(2) Standard curve method If the pathlength b is constant, for the same absorbing substance, Beer-Lambert law could be written:

$$c \propto A$$

Hence, the absorbance of a series of standard solutions detected by spectrophotometer could be plotted as ordinate and against the concentration as abscissa. The standard curve is established in Figure 8. 6 and should pass through the point with $A = 0$ and $c = 0$. The concentration of the test substance could be determined on the standard curve depending on the absorbance of the tested solution. The standard curve method is relatively convenient, with little error and large numbers of sample detections.

The features of spectrophotometry include: ①The range of the straight-line portion in the standard curve is wider, and the accuracy of the analysis result is higher since the incident light, that is, the monochromatic light with higher purity, could result in less deviation of Beer-Lambert law. ②The monochromatic light of a certain wavelength could be selected, and two or more solution components could be detected simultaneously by the additivity of absorbance. ③Since the wavelength range of the incident light is extended, lots of colorless substances absorbed in the ultraviolet or infrared region also could be detected by absorption spectrophotometry.

3. Spectrophotometer

According to the scope of the wavelength range, a spectrophotometer with a measuring wavelength of 420-700 nm is called a visible spectrophotometer, and the wavelength of a spectrophotometer with a measuring wavelength of 360-800 nm is called an ultraviolet-visible spectrophotometer. A wavelength of 25 000 nm is measured using an infrared spectrophotometer. Visible and ultraviolet-visible spectrophotometers are mainly applied to the content detection of inorganic and organic materials, while infrared spectrophotometers are mainly used in structural analysis. Common spectrophotometers are 721 type and 722 type spectrophotometers. Furthermore, spectrophotometers could also fall into two categories: single-beam and double-beam.

The basic parts of a spectrophotometer include a light source, a monochromator, a sample cell, a detector, and a monitor, as shown in Figure 8. 7.

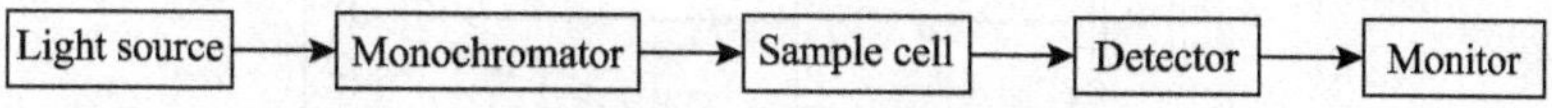

Figure 8. 7 The main device of the spectrophotometer

(1) Light source A light source should provide a beam of radiation with a certain wavelength that has stable power and is easy to detect. The early light source was the tungsten lamp, which had wide wavelengths (320-2 500 nm). To accommodate the higher temperature, the tungsten/halogen lamp containing a certain amount of iodine covered with the quartz occurs. The tungsten/halogen lamp (also quartz/halogen lamp) has the same wavelength range but a higher adaptive temperature and a longer lifetime than the tungsten lamp. Thus, the tungsten/halogen lamp shows increasing use in ultraviolet, visible, and near-infrared regions. The hydrogen or the deuterium lamp could be used for the far-ultraviolet ($<$200 nm) detection. Moreover, the xenon lamp with a spectral range of 250-600 nm could also be used, although the stability is poor.

(2) Monochromator A monochromator is an optical device that can disperse the

polychromatic light excited from the light source to the monochromatic light. According to the sequence of wavelengths, white light disperses the monochromatic light by passing through a prism or a grating. A monochromator comprises five parts: entrance slit, collimator, dispersion, projector, and exit slit.

The disperser is the crucial part of the monochromator, and it mainly includes two types: the prism and the diffraction grating.

Common prisms have glass prisms, quartz prisms, or fused silica prisms. The glass prism with high dispersion capacity could absorb the visible region light between 400 and 1 000 nm. On the other hand, the quartz prism or the fused silica prism is required in the ultraviolet light region below 400 nm.

Diffraction gratings are the dispersion device by the principle of diffraction from a certain number of closely spaced lines and equally wide marks on a surface. The grating features are uniform dispersion, linear relation, automation of photometric measurement, and wide band.

In addition to the disperser, an entrance slit could reduce the incident beam from the light source, and an exit slit could select the light with a certain wavelength to radiate into the text solution. The collimator, usually made of a lens or a concave mirror, can produce a collimated beam from the incident light. The projector is generally a lens or a concave mirror that could do the monochromatic light project to the exit slit.

The optical system principle of the monochromator in the grating spectrophotometer is shown in Figure 8. 8.

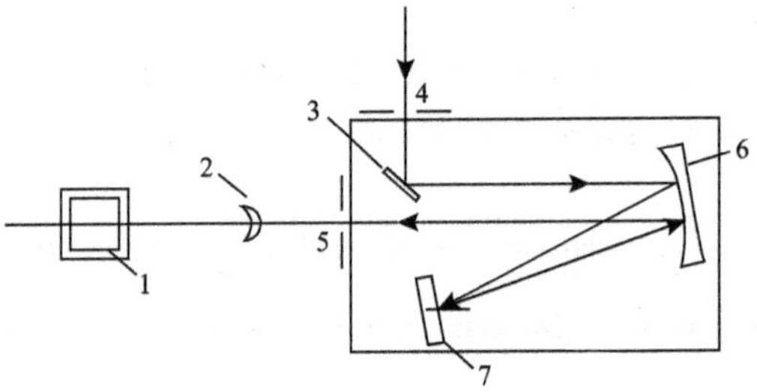

1. Sample cell 2. Condenser 3. Plane mirror 4. Entrance slit
5. Exit slit 6. Spherical mirror 7. Plane grating

Figure 8. 8 The schematic diagram of the optical system of the monochromator

From Figure 8. 8, the entrance slit and the exit slit are placed on the focal plane of spherical mirrors. The light from the slit reflected from the plane mirror to the spherical mirror is collimated, and the collimated light is reflected from the spherical mirror to the plane grating. Then the collimated light diffracted from the grating is reflected on the exit slit by the spherical mirror.

The monochromatic light from the slit of the monochromator is a band with a certain spectral width, named spectral bandwidth. The width of the slit in the monochromator could be adjusted. The narrower width of the slit could result in a better resolution of the absorption band. However, the narrower bandwidth could also reduce the intensity of the photocurrent signal.

(3) Sample cell The sample cell contains the cell (or cuvette) and the cell carrier. The text solution in the cell should be placed in the cell carrier to locate the radiation beam. The cell is made of glass or quartz glass with different thicknesses. The glass cell could only be used in the visible light region, while the quartz glass cell could be used in the visible light but ultraviolet regions. The commonly used cell is the cuvette with 1 cm in path length (Figure 8. 9). It should be noted that the cell should be kept clean, and the transmit side should not wear out during use.

(4) Detector The detector is a photoelectric transformation device that transforms the light's signal intensity from the absorption cell to the measurable signal intensity of the electricity for measurement. The photoelectric and photomultiplier tubes are commonly used in visible-ultraviolet spectrophotometers.

The photoelectric tube is a vacuum diode, the anode is a wire, the cathode is a semicircular cylinder of metals, and the inside is coated with photographic material. According to the different performances, photographic material falls into two basic types: infrared sensitivity and ultraviolet sensitivity. The cathode surface of the infrared-sensitive photoelectric tube is coated with silver and cesium oxide, and its applied range is 625-1 000 nm. The cathode surface of the ultraviolet-sensitive photoelectric tube is coated with antimony and cesium, and its applied range is 200-625 nm.

The photomultiplier tube is a vacuum photographic device that uses photocurrent from secondary electron emission. It comprises a photoelectric emission cathode, an anode, and several multiplication electrodes.

(5) Monitor The early spectrophotometer monitor commonly adopted the galvanometer or the microammeter to measure the output of the detector, which could directly display the absorbance or transmittance. The modern spectrophotometer commonly uses the digital voltmeter instead, which can display the absorption spectrum on the screen. The whole procedure might be automated.

4. Instrument

Take the 722-type spectrophotometer as an example to illuminate the application method (Figure 8. 9).

Check whether the starting position of the instrument adjusting knob is correct. Next, turn on the power switch and open the cassette of the sample chamber. Mark the electricity meter pointer read "0" and preheat it for 20 min. Then, choose the monochromatic light wavelength needed, and the corresponding magnifies the sensitivity button, adjust the electricity meter to $T=0$ by the "0" transmittance regulator.

Close the sample chamber to make the phototube accept the light, and push the handle of the sample holder to place the reference solution cell on the light path. Adjust the "100%" transmittance regulator to make the electricity meter pointer read "$T=100$".

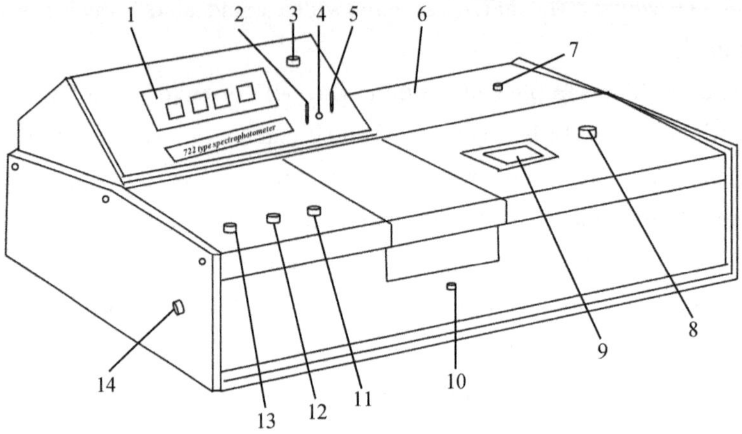

1. Digital display 2. Absorbance zero adjustment knob 3. Select switch
4. Absorbance adjustable slope potentiometer 5. Concentration knob 6. Light source room
7. Power switch 8. Wavelength handwheel 9. Wavelength scale window 10. Sample holder handle
11. T-100% knob 12. T-0 knob 13. Sensitivity adjustment knob 14. Dryer

Figure 8.9 722-type spectrophotometer

Repeat the process that opens the sample chamber to adjust "0" and close the chamber to adjust the transmittance to 100% until the instrument remains stable.

Close the sample chamber and push the handle of the sample holder. Next, make the sample solution cell on the light path and read the absorbance value. The sample chamber should be opened immediately after obtaining the absorbance.

Please remove the absorption cell after the measurement, clean it, and invert it to the filter to dry. Next, turn all buttons to the original position, turn off the power switch, and pull out the attaching plug.

The sensitivity of all the grades of the amplifier is "1"×1 times, "2"×10 times, and "3"×20 times, which the sensitivity increases successively. Since the light energy with different monochromatic light wavelengths is different, the different sensitivity grades should be chosen. The principle of choice is that the lower sensitivity grade should improve the instrument's stability when the reference solution is adjusted to $T = 100\%$. The electricity meter should be adjusted to 0 and 100 as the sensitivity grade changes.

8.6 Condition Choice of Spectrophotometer

1. The choice of the test wavelength

Two principles should be followed when selecting the appropriate incident light wavelength: The principle of maximum absorption. The maximum absorption wavelength light of the test material should be taken as the incident light (no interference at the maximum absorption wavelength). And the principle of maximum absorption and minimum interference. When selecting a wavelength to measure material interference, choosing one that exhibits peak absorption and minimal interference is crucial. This ensures that the interference from other materials is gauged accurately at the wavelength of maximum absorption. However, it is essential to note that the wavelength with the highest absorption should not be chosen if it leads to increased interference, thereby compromising the precision of the measurement.

2. The choice of the absorbance scope

As shown in Figure 8.10, the measurement of the relative error is smaller, and the accuracy is higher as the transmittance T is at 0.15-0.65 (the corresponding absorbance $A = 0.2$-0.8). The following methods could be used to control the scope of the absorbance if beyond the scope: change the quality of the test sample and dilute the solution to control the concentration of the solution. Select the absorption cell with different thicknesses.

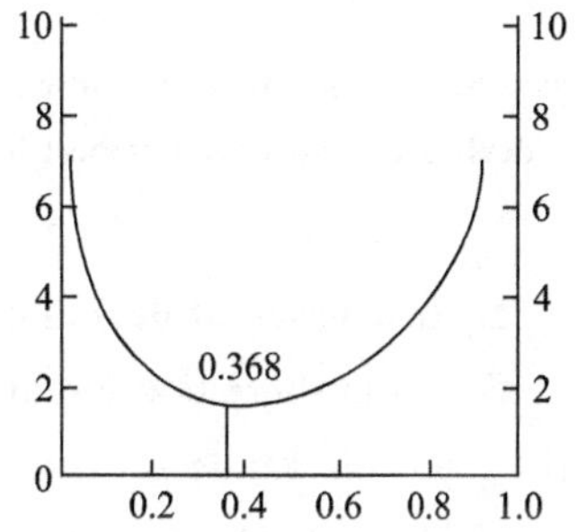

Figure 8.10 The relation between transmittance and relative error

3. The choice of the reference solution

The zero of the instruments could be adjusted by the reference solution so that the error is taken from the absorption cell and the solvent to the reflection, the absorption of the incident light could be eliminated, and the influence of the interference could be avoided. The reference solution should be selected depending on the specific condition of the sample solution and the color agent.

The distilled water should be the reference solution if the sample solution and the color agent are colorless. Likewise, suppose the color agent is colorless, and the other colored ion exists in the test sample. In that case, the test sample without the color agent should be the reference solution.

The reagent without the color agent should be selected as the reference solution if the color agent has color.

If the color agent and the test sample both have color, the appropriate masking agent could be added to the test sample so it does not react to the color agent. In addition, the color agent and other reagents could be added to the solution according to the test sample measurement, which could be taken as the reference solution to eliminate the interference of both the color agent and the coexisted component.

Change the order of adding reagents to ensure the test component does not react with the color, which could be taken as the reference solution to eliminate the interference.

8.7 Error of Spectrophotometry

8.7.1 Error Deviated from Beer-Lambert Law

When using the standard curve method to measure the unknown concentration of the

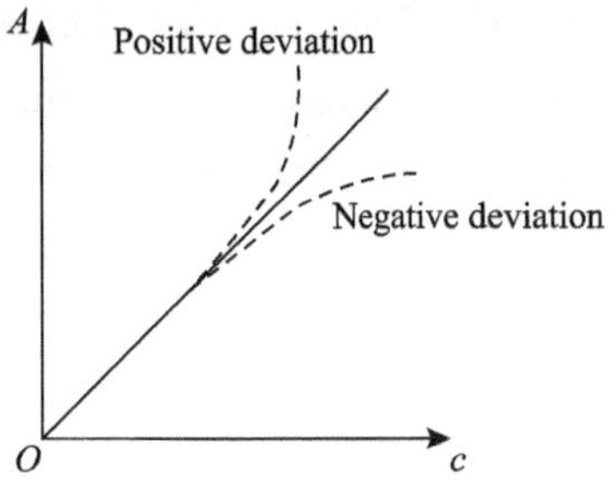

Figure 8. 11 The standard curve and the deviation from Beer-Lambert law

solution, the standard curve generally is a straight line passing through the origin, which is consistent with Beer-Lambert law. However, some exceptions could bend the standard curve from the origin to the concentration or absorbance axes. This might be called **deviation from Beer-Lambert law** (Figure 8.11). The quantitative measurement in the part of curvature could cause a larger error. The deviation from Beer-Lambert law is that the actual condition of the instrument or the solution differs from the ideal condition of Beer-Lambert law.

1. Instrumental deviation: polychromatic light

Beer-Lambert law could only be used in monochromatic light. However, the incident light from all kinds of spectrophotometers is the polychromatic light with a certain band caused by the restriction of the dispersion capacity of the monochromatic light and the certain wide kept from the exit slit. The different absorption degrees from the substance to the different wavelength lights could result in a deviation from the Beer-Lambert law.

The polychromatic light is made of λ_1 and λ_2. The absorbance deviation caused by the polychromatic light of solutions "a" and "b" with different concentrations may differ. The greater the density, the greater the error from polychromatic light, so the linear relationship will gradually bend downward as the density increases.

Measures to avoid the deviation from the non-monochromatic light are as follows:

A better monochromator could be used to obtain the highly purified monochromatic light and the wide linear range of the standard curve.

The maximum absorption wavelength of the test material could be selected as the wavelength of the incident light, which could ensure the high sensitivity of the measurement. Furthermore, since the deviation from the non-monochromatic light at the maximum absorption wavelength is smaller than at other wavelengths, the absorption curve is relatively flat, and ε of the light from each wavelength near the maximum absorption wavelength is about equal in value.

The appropriate concentration range should be selected to make the absorbance reading in the linear range of the standard curve.

2. Heterogeneous solution deviation

According to the Beer-Lambert law, the solution of the absorption material should be homogeneous. Suppose the test solution is heterogeneously included in the colloidal solution. In that case, the emulsion and the suspension, a part of the incident light could be absorbed, and the other part of the incident light could be lost by scattering when passing through the solution. The transmittance decreases, and the measured absorbance increases, so the standard curve deviates from the straight line to the absorbance axis. The solution should be avoided to generate the colloidal or turbidity in absorption spectrophotometry.

3. Chemical deviation

The concentration of the absorption substance in the solution could be caused by chemical changes, including dissociation, association, production, or tautomerization, which could lead to the deviation from Beer-Lambert law. When the concentration of the absorption substance is high enough to 0.01 mol · L^{-1}, the average distance between particles is reduced, and the electrostatic interactions are increased, which leads to deviation from Beer-Lambert law.

(1) Dissociation The acid and alkali forms of most organic acid and alkali compounds have different absorption properties for light. Therefore, the different acidity and the different acid (alkali) dissociation of the solution could result in a change in the ratio between the acid and the alkali, and the absorbance of the solution also changes.

(2) Complexation The multistage complex generated by the color agent with the metal ion has different absorption properties for light. For example, in the complex of Fe (Ⅲ) with SCN^-, the color of $Fe(SCN)_3$ is deepest, while $Fe(SCN)_2$ is lightest, so the larger concentration of SCN^- could result in the deeper color of the solution, which the absorbance could also be increased.

(3) Association For instance, CrO_4^{2-} could be associated with $Cr_2O_7^{2-}$ in the acidic condition, while the two complexes have different absorbances to light.

The condition of the solution should be controlled to keep the test sample in one form that could overcome the deviation from the Beer-Lambert law caused by the chemical factors.

8.7.2 Absorbance Error in Measurements

As shown in Figure 8.10, since transmittance has a logarithmic relationship with concentration, small errors in measuring transmittance could cause large relative errors in the calculated concentration. The transmittance is lower or higher. Therefore, the relative error is larger. Thus, choosing the transmittance reading in the middle of the scale is better.

To keep a smaller relative error, the transmittance of the test solution could occur in the range of 5%-15% ($A=0.2$-0.8) when $T=36.8\%$ ($A=0.434$), the relative error of the measurement is minimum.

Hence, the test solution should be diluted or concentrated, ensuring the optimum absorbance range.

8.8 Other Absorption Spectrophotometry

8.8.1 Differential Spectrophotometry

Absorption spectrophotometry could only be used in micro-component measurement. The higher or lower concentration of test components might cause a large measurement error, leading to decreased accuracy. **Differential spectrophotometry** could overcome this shortcoming. Differential spectrophotometry mainly includes high-concentration differential spectrophotometry, low-concentration differential spectrophotometry, and precision differential spectrophotometry with two reference solutions. Their basic principles are the same, but the high concentration differential spectrophotometry is now widely used. The following text only introduces the high concentration differential spectrophotometry.

Compared with the test solution with an unknown concentration, the standard solution with a known concentration, the reference solution, could overcome the larger error caused by the higher or lower concentration of the test component with common spectrophotometry.

The principle: the concentration of the standard solution is c_s, the test solution is c_x, $c_x > c_s$,

$$A_x = \varepsilon b c_x \qquad A_s = \varepsilon b c_s$$

Subtraction of the above formulas: $\Delta A = A_x - A_s = \varepsilon b (c_x - c_s)$

$$\Delta A = \varepsilon b \Delta c$$

The measured absorbance equals the absorbance difference ΔA between the test solution and the standard solution in the common method. The standard counting and curve methods could be used in quantitative analysis depending on the principle.

(1) Standard counting method Two standard solutions c_1 and c_2 are prepared ($c_1 < c_2$). The absorbance of the c_2 and the c_x (A_2, A_x) are measured respectively by taking c_1 as the reference solution, and then the c_x could be obtained.

(2) Standard curve method Initially, the standard series solutions are prepared. Then, the respective ΔA and the test solution ΔA_x are measured by taking the standard solution with the minimum concentration as the reference. Finally, the standard curve of $\Delta A - \Delta c$ could be pointed, and the concentration of the test solution could be determined.

Differential spectrophotometry could improve the method's accuracy by improving the measurement's accuracy. The transmittance of the reference solution is close to the test component, so the error of the measured relative concentration is smaller.

8.8.2 Dual Wavelength Spectrophotometry

The principle of **dual-wavelength spectrophotometry** is that two beams of monochromatic light with different wavelengths irradiate the same absorption cell at a certain time interval. Then, the difference between the two absorbances can be measured and recorded (Figure 8.12). The signal of the reference wavelength from the signal of the analysis wavelength is deducted so that the content of the test component can be obtained by eliminating interference. The sensitivity, selectivity, and accuracy of dual-wavelength spectrophotometry are high and widely used to analyze environmental and biological samples.

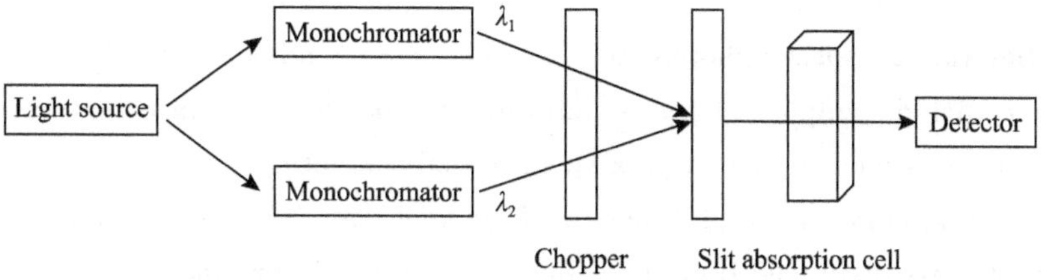

Figure 8. 12 The diagram of dual-wavelength spectrophotometry

$$\Delta A = A_{\lambda_1} - A_{\lambda_2} = (\varepsilon_{\lambda_1} - \varepsilon_{\lambda_2})bc$$

The theoretical basis of quantitation is that ΔA is proportional to the concentration of the absorption material. Therefore, taking the absorbance of the test solution for a certain wavelength of light as the reference only by one absorption cell could eliminate the measurement error from the difference between the test solution with the reference solution and between two absorption cells so that the measurement accuracy could be improved.

Dual-wavelength spectrophotometry is mainly used in the component measurement of the turbid solution, single component, and the respective measurement with two coexisting components.

(1) The component measurement of the turbid solution The maximum absorption wavelength of the test component is commonly selected as the measurement wavelength (λ_1). The wavelength is close to a 40-60 nm difference, and a large A value is selected as the reference wavelength.

(2) The single-component measurement The absorption peak of the complex is taken as the measurement wavelength in single component measurement. The reference wavelength includes the absorptive point wavelength, the certain wavelength with the lower part on the absorption curve of the colored complex, and the absorption peak of the color agent.

(3) The respective measurement with two coexisting components Since the absorption spectrometry of two components overlaps, the optical absorption of the other component should be eliminated to measure the one component. For the two-component systems interfered with each other, their absorption spectrometry overlaps so that the condition of the selected reference wavelength and the measured wavelength are described as follows: The absorbance difference ΔA of the test component at two wavelengths is large enough. In comparison, the absorbances of the interference component at two wavelengths are equivalent. Then, the absorbance difference measured by dual-wavelength absorbance is only linearly related to the concentration of the test component, which has nothing to do with the interference component. Thus, the interference could be eliminated.

8. 8. 3 Derivative Spectrophotometry

Since the absorption result could provide more information from derivative plots than from the absorption curve, derivative spectrophotometry has received increasing attention. **Derivative spectrophotometry** shows great advantages in multicomponent measurement, turbid samples analysis, eliminating background interferences, reinforcing the spectrum's fine structures, complex spectra analysis, etc.

An absorption curve is as follows in Figure 8.13. The first-order and third-order derivative spectra are dispersed functions that differ from the original curve. Therefore, they can be used to revise the λ_{max} of the absorption curve accurately. The second and fourth-order derivative spectra have a strong peak with the same height but are sharper than the original curve. Just like these, the resolution of the even-order derivative spectra is gradually improved so that they can be used to separate two absorption bands, the mixture band in the original curve.

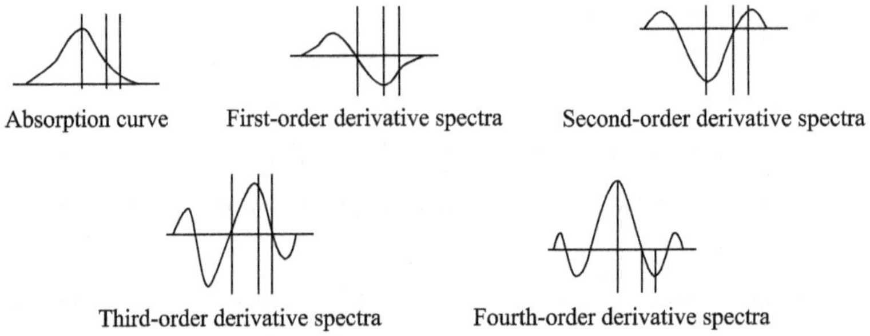

Figure 8.13 The derivative curve of the absorbance to the wavelength

With the development of electronics, more spectrophotometers could deal with the higher-order derivative spectra to solve intractable analytical problems.

8.9 Application of Absorbance

Absorbance has broad applications, and it's mainly used in the measurement of micro-components, the measurement of the composition of the complex and stability constant; the rate constant of the reaction, the activation energy of the catalyzed reaction, etc.; deducing the spatial configuration of the molecule depended on the ultraviolet or infrared spectrum data, determining the molecular structure, etc. This book describes examples only for the micro-component measurement of the test sample.

1. The measurement of trace Fe

1,10-phenanthroline could react with Fe^{2+} to form the red-colored complex ion with 3 : 1 at pH=3-9 (commonly at pH=5). $\lambda_{max}=512$ nm, $\varepsilon=1.1\times10^4$ L · mol^{-1} · cm^{-1}. The reaction could complete quickly and keep a stable pH of 3-9 color. The concentration and the absorbance are consistent with the Beer-Lambert law when the content of Fe is 0.5-8 μg · L^{-1}. The buffer with pH=4.5-5.0 could be used to keep the acidity, and the hydrochloric acid could be used to reduce Fe^{3+} so that the trace Fe could be measured by the standard curve method. This method has high accuracy, good reproducibility, and high selectivity.

2. The measurement of phosphorus

The sample could react with concentrated sulphuric acid and perchloric acid to make all the forms of phosphorus turn to H_3PO_4. Then, H_3PO_4 reacts with $(NH_4)_2MoO_4$ to form phosphorus molybdenum yellow heteropolyacid in an HNO_3 solution. The reaction is as follows:

$$PO_4^{3-} + 12MoO_4^{2-} + 27H^+ = H_7P(Mo_2O_7)_6 + 10H_2O$$

The appropriate reductant (such as vitamin C) could reduce Mo(Ⅵ) to Mo(Ⅴ), which generates the blue-colored phosphomolybdenum blue. The maximum absorption wavelength of phosphomolybdenum blue is $\lambda_{max} = 650$ nm, and it's consistent with the Beer-Lambert law in the phosphorus content range 0.05-2.0 $\mu g \cdot L^{-1}$, so the standard curve could measure the phosphorus content of the test sample. The common reductant of the phpsphomolybdenum blue method includes stannic chloride and ascorbic acid. Despite the high sensitivity and the quick development by using stannic chloride, the stability of the blue color is bad, and the acidity and the concentration of ammonium molybdate are required strictly. The blue color is stable, and the acidity range required is wide without interference using ascorbic acid as the reductant. Still, the rate of color reaction is slow, and a boiling water bath is needed to heat.

Key Glossary

Absorption spectrophotometry: Absorption spectrophotometry is an analytical technique used to measure the concentration of a substance in a solution based on the amount of light the substance absorbs. By analyzing the light intensity before and after it passes through the sample, the technique quantifies how much of the light is absorbed at specific wavelengths, which is directly related to the concentration of the absorbing species in the sample.

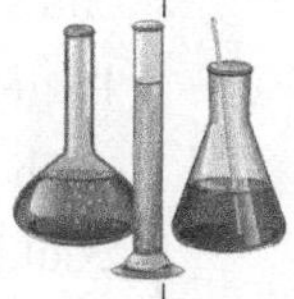

Polychromatic light: Light is composed of rays with different wavelengths.

Monochromatic light: The light with a single wavelength.

Complementary color light: Suppose the white light is a mixture with a certain proportion of two different colored monochromatic lights. In that case, the pair of colored monochromatic lights are called complementary color light, which is complementary to light.

Absorption spectra curve (or **Absorption curve**): Accurately express the absorption degree of the solution with different wavelengths of light, generally taking the absorption degree(absorbance) as the Y-axis and the wavelength as the X-axis. The curve could be established.

The law of light absorption (or **Beer-Lambert law**): The absorbance of the solution is directly proportional to its concentration and the optical path (thickness of solution) when a parallel monochromatic light irradiates the sample solution vertically.

Absorption coefficient: The absorbance of absorption material in unit concentration and unit thickness. Its unit is $L \cdot g^{-1} \cdot cm^{-1}$.

Molar absorption coefficient: The absorbance of absorption material in unit concentration and unit thickness. Its unit is $L \cdot mol^{-1} \cdot cm^{-1}$.

Color reaction: If the color of the test substance is colorless or light-colored, the appropriate reagent is needed to react with test ions to produce a colored compound. The reagent is called a color reagent.

Visual colorimetry: An analytical method could determine the material content by comparing the color depth of the solution with the eyes.

Spectrophotometry: An analytical method that could measure the absorbance of the solution by spectrophotometer and determine the concentration of the test solution depends on the Beer-Lambert law.

Deviation from the Beer-Lambert law: The standard curve would be away from the linear relationship in absorption spectrophotometry, and the curve would mainly be bent from the origin to the concentration axis with the higher concentration of absorbing materials.

Differential spectrophotometry: Absorption spectrophotometry can be measured in higher or lower concentrations of test components. The principle is compared to the unknown concentration of the test solution with the known concentration of the standard solution, which is the reference solution that could overcome the larger error caused by the higher or lower concentration of the test component with common spectrophotometry. The test solution of unknown concentration could be measured.

Dual-wavelength spectrophotometry: In the absorption spectrophotometry method, the principle is that two beams of monochromatic light with different wavelengths alternately irradiate the same absorption cell at a certain time interval. Then, the difference between the two absorbances could be measured and recorded.

Derivative absorption spectrophotometry: The principle of derivative absorption spectrophotometry is the measurement of the derivative curve by absorbance (or transmittance) for wavelength. Thus, depending on the derivative curve, the concentration of the test sample could be measured.

Essential Exercises

Review Questions

1. Explain the physical meaning of Beer-Lambert law. Why could this law only be used in monochromatic light?

2. Describe the main device, categories, and the basic performance of the spectrophotometer.

3. Illustrate why the concentration of test solutions should be measured at $\lambda_{max.}$ For example, what is the selection principle of the wavelength of incident light?

4. Why should we choose an appropriate reference solution? How do you select the reference solution?

Cumulative Problems

1. The concentration of a solute is 5×10^{-3} g · L^{-1} in the solution, and its molecular weight is 125 g · mol^{-1}. This solution is put in the absorption cell whose thickness is 2.0 cm and the absorbance is 2.00. Calculate the a and ε of this solute.

2. Weight accurately 0.200 g vitamin C and dissolve it in 100 mL 0.005 mol · L^{-1} sulfuric acid solution. Take the solution in the 1 cm absorption cell, and the absorbance is 0.616 at $\lambda_{max}=245$ nm. Next, calculate the composition of vitamin C in the sample.

3. A beam of monochromatic light goes through the colored solution, the thickness of which is 1.0 cm, and the transmittance is 80%. If the thickness of the colored solution is 5.0 cm, calculate the transmittance under the same conditions.

4. Detect SiO_2 by silicon molybdenum blue spectrophotometry. Take the SiO_2 standard solution with a concentration of 0.010 mg · mL^{-1} as the reference solution, and detect another SiO_2 standard solution with a concentration of 0.100 mg · mL^{-1}. The transmittance is 14.4%. An unknown solution's transmittance is 31.8% in the same condition. Calculate the SiO_2 concentration of this unknown solution.

Chapter 9

Potentiometric Analysis

Key Points

- *Potentiometry*
- *pH Electrode*
- *Ion-selective Electrode*
- *Indicator Electrode and Reference Electrode*

9. 1 Basic Principle of Potentiometry

Electrochemical analysis uses electrochemistry to analyze chemicals. There are many types of methods for electrochemical research, including potentiometry, amperometry, voltammetry, and coulometry. In this chapter, the technique of potentiometry is introduced.

Potentiometry is an electrochemical analysis technique measuring a cell potential with zero current passing through the system. There are two types of potentiometry: One is potentiometric titration, which utilizes a potential measurement in a titration with various amounts of titrant in the analyte. The other type is the direct potentiometric method.

9. 1. 1 Nernst Equation and Cell Potential

The Nernst equation (Equation 9. 1) illustrates the relationship between the reactant concentrations and electrode potential.

For a general half-reaction: $Ox + ne^- \rightleftharpoons Red$, the Nernst equation is:

$$\varphi=\varphi^{\ominus'}+\frac{RT}{nF}\ln\frac{c_{Ox}}{c_{Red}} \qquad \textbf{(Equation 9. 1)}$$

Chemical analysis by potentiometry will measure the potential difference between **an indicator electrode** and **a reference electrode. The indicator electrode** is the cathode electrode in contact with the sample and gives a potential related to the activity and concentration of the analyte. **The reference electrode**, initially set as the anode in an electrochemical cell, provides a fixed potential against which the potential of the indicator electrode can be measured. A salt bridge is used to connect the two half-cells with separated contents.

The Nernst equation calculates the expected potentials at the cathode and anode in an electrochemical cell. The potential difference between these electrodes will be measured by

Equation 9. 2 if essentially no current exists.

$$E_{Cell}=E_{Cathode}-E_{Anode} \quad \textbf{(Equation 9. 2)}$$

The potential difference between the two electrodes is measured by potentiometry in an electrochemical cell. Equation 9. 2 can thus be written in the following form:

$$E_{Cell}=E_{Ind}-E_{Ref} \quad \textbf{(Equation 9. 3)}$$

$$E_{Ind}=E_{Cell}+E_{Ref} \quad \textbf{(Equation 9. 4)}$$

where E_{Cell} is the potential difference measured between these two electrodes. E_{Ind} is the potential of the indicator electrode, and E_{Ref} is the potential of the reference electrode.

If the indicator electrode is set as the anode, the equation can be rewritten as below:

$$E_{Cell}=E_{Ref}-E_{Ind} \quad \textbf{(Equation 9. 5)}$$

$$E_{Ind}=E_{Ref}-E_{Cell} \quad \textbf{(Equation 9. 6)}$$

9. 1. 2 Indicator Electrode

Many types of indicator electrodes can be used for potentiometry. Such as:

1. Metallic indicator electrode

The metallic indicator electrode is the first generation of electrodes used in potentiometric analysis and is still widely used today. It includes three types:

The first type is the **inert metal electrode**, which uses an inert metal as an electrode to oxidize or reduce another substance. It is made from a material such as platinum, palladium, or gold. The inert metal electrode does not participate in the reaction but only provides a place for electronic exchange. For example, a platinum electrode acts as a cathode to transform Fe^{3+} to Fe^{2+}. The half-reaction and Nernst equation at 25 ℃ are given as:

The half-reaction: $Fe^{3+} + e^- \rightleftharpoons Fe^{2+}$

The Nernst equation:

$$\varphi=\varphi^{\ominus}(Fe^{3+}/Fe^{2+})+0.059\,2\lg\frac{c(Fe^{3+})}{c(Fe^{2+})}$$

The section type is the **active metal electrode** consisting of a metal in contact with a solution containing the same metal ions. An example of this electrode is a silver wire immersed in a silver nitrate solution. The potential of this electrode depends on the activity and concentration of Ag^+ in the solution. The half-reaction and Nernst equation are given as:

The half-reaction: $Ag^+ + e^- \rightleftharpoons Ag$

The Nernst equation: $\varphi=\varphi^{\ominus}+\frac{RT}{F}\ln c(Ag^+)$

The third type is the **metal-slightly soluble metallic salt electrode**, which consists of metal in a solution containing the anion of a slightly soluble salt of that metal. An example of this electrode is a silver wire in contact with AgCl(s) immersed in a solution containing chloride ions (the anion that reacts with Ag^+ to form solid AgCl). The half-reaction and Nernst equation at 25 ℃ are given as:

The half-reaction: $AgCl(s) + e^- \rightleftharpoons Ag^+ + Cl^-$

The Nernst equation: $\varphi(AgCl/Ag) = \varphi^\ominus(AgCl/Ag) + 0.059\ \lg \frac{1}{c(Cl^-)}$

2. Membrane electrode

Ion-selective electrodes and pH glass-membrane electrodes are both membrane electrodes.

9.1.3 Reference Electrode

A reference electrode is to provide a stable potential that is reproducible and easy to prepare. For example, **a saturated calomel electrode** is a common reference electrode that is stable and reproducible. For the calomel electrode, an insoluble chloride salt coats the salt's free element (Hg), and free elements are immersed in a KCl solution of known concentration (often saturated KCl). A calomel electrode containing a saturated KCl solution is known as a saturated calomel electrode (SCE). The half-reaction and Nernst equation at 25 ℃ are given below.

The half-reaction: $Hg_2Cl_2(s) + 2e \rightleftharpoons 2Hg(s) + 2Cl^-$

The Nernst equation: $\varphi(Hg_2Cl_2/Hg) = \varphi^\ominus(Hg_2Cl_2/Hg) - 0.0592\ \lg c(Cl^-)$

According to the Nernst equation, the potential of a calomel electrode varies with the concentration of chloride ions at a specific temperature. Table 9.1 represents the potential values at different concentrations of KCl at 25 ℃.

Table 9.1 The values of potential at different concentrations of KCl

Concentrations of KCl	$0.1\ mol \cdot L^{-1}$	$1\ mol \cdot L^{-1}$	Saturated KCl
Potential/V	+ 0.336 0	+ 0.208 4	+ 0.243 8

The potential of a calomel electrode also varies with the temperature. However, the calomel electrode becomes unstable at a temperature above 80 ℃. One other type of reference electrode is the silver-silver chloride electrode.

9.2 Potentiometric Analysis of pH

9.2.1 Glass Membrane Electrode

A glass-membrane electrode is the most common hydrogen ion-selective electrode, which can be used to measure the pH of all kinds of solutions conveniently. It is an indicator electrode that uses a thin glass membrane to detect the desired ion selectively. The glass used in a glass-membrane electrode consists of 22% Na_2O, 6% CaO, and 72% SiO_2. The pH value is slightly less than 9, so the glass can respond well to the concentration of H^+. The design of a glass-membrane electrode is given in Figure 9.1. The thickness of the lower hemisphere of the glass ball ranges from 0.05 mm to 0.15 mm. The ball's HCl solution (0.1 mol • L^{-1}) is the internal reference solution. The silver-silver chloride electrode immersed in an internal reference solution is the inner reference electrode.

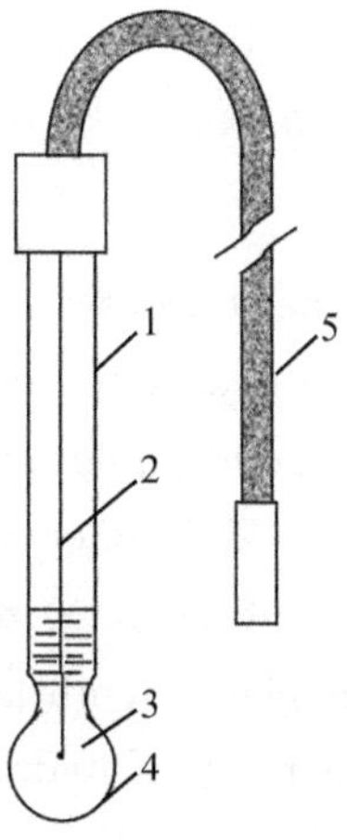

1. Glass tube 2. The inner reference electrode (Ag/AgCl) 3. Inner reference solution (0.1 mol • L^{-1} HCl) 4. Glass-membrane 5. Wire

Figure 9.1 Design of a glass-membrane electrode

The glass membrane is between an internal reference solution (fixed hydrogen ion activity) and a sample solution during the test. A junction potential is presented when two solutions or regions exist in an electrochemical cell with different chemical compositions. For example, the inner glass membrane's internal reference solution and hydration layer and the outer glass membrane's sample solution and hydration layer have different chemical compositions. Therefore, the boundary phase potentials (φ_{Inn} and φ_{Out}) exist. The membrane potential is formed by ion exchange and diffusion. The difference in boundary phase potentials is called the membrane potential (φ_{Mem}), which is related to the hydrogen ion activity of the external glass membrane. The relationship is given as:

$$\varphi_{Mem} = \varphi_{Out} - \varphi_{Inn} = 0.0592 \lg \frac{\alpha_{H^+(s)}}{\alpha_{H^+(Inn)}} \quad \textbf{(Equation 9.7)}$$

If the hydrogen ion activity of the internal reference solution is fixed, Equation 9.7 can be rewritten as:

$$\varphi_{Ele} = K + 0.0592 \lg \alpha_{H^+(s)} = K - 0.0592 \text{ pH} \quad \textbf{(Equation 9.8)}$$

where K is a constant relating to the property of a pH electrode at a given temperature, the membrane potential and the pH of the sample solution are linearly related.

The specific glass in a glass-membrane electrode consists of a silicate lattice of a negative charge. In contrast, the sodium ions that are small and active in the crystal lattice can conduct electricity. Therefore, when the glass ball is immersed in water, a thin hydration layer will be formed on the surface. Furthermore, the bond between the silicate and hydrogen ion is stronger than the sodium ion (about 10^{14} times), resulting in an ion-exchange reaction:

$$H^+ + Na^+Cl^- \rightleftharpoons Na^+ + H^+Cl^-$$

If the hydration layer contacts the sample solution, the hydrogen ions in the hydration layer and sample solution will inter-exchange. The equilibrium is established on the surfaces of the internal and external hydration layers.

$$H^+_{glass} \rightleftharpoons H^+_{solution}$$

However, the concentration of hydrogen ions is different. Extra hydrogen ions can thus enter or leave the hydration layer from the sample solution. As a result, the distribution of charges on the solid-liquid interface is changed. The junction potential of the outer membrane (φ_{Out}) and inner membrane (φ_{Inn}) is then formed. If the concentration of hydrogen ions in the inner membrane is fixed, the potential of the inner membrane is also fixed. Thus, the glass-membrane potential is related to the potential of the outer membrane, influencing the activity of hydrogen ions in the sample.

If $\alpha_{H^+(s)} = \alpha_{H^+(Inn)}$, then $\varphi_{mem} = 0$. The glass membrane potential is not exactly 0 mV, which is called the **asymmetric potential** (φ_{Asy}). It forms due to different outer and inner membrane surface properties, e.g., constitution, surface tension, the extent of hydration, exchange capacity, etc. This potential ranges from 1 to 30 mV. The glass-membrane electrode exhibits similar asymmetric potential, which is constant.

Before the test, the glass-membrane electrode must be kept in pure water for at least 24 hours to create a hydration layer of sensitive hydrogen ions and decrease the asymmetric potential. This is because a dry glass membrane is unstable for the change of hydrogen ions. The initial asymmetric potential has the highest value for an immersed glass-membrane electrode, decreasing gradually to a fixed value in water.

In summary, the potential of the glass-membrane electrode includes the inner reference electrode potential (φ_{Inn}), membrane potential (φ_{Mem}), and asymmetric potential (φ_{Asy}).

For example:

AgCl | 0.1 mol · L^{-1} HCl | glass-membrane | sample solution

Potential of inner reference electrode (AgCl | 0.1 mol · L^{-1} HCl)   Membrane potential (glass-membrane | sample solution)

The potential of the glass-membrane electrode is:

$$\varphi_{glass} = \varphi_{Inn} + \varphi_{Mem} + \varphi_{Asy} = \varphi_{Inn} + K - 0.0592\ pH + \varphi_{Asy} \qquad \textbf{(Equation 9.9)}$$

If $\varphi_{Inn} + K + \varphi_{Asy} = K'$, $\varphi_{glass} = K' - 0.0592\ pH$.

The glass membrane's potential and the sample solution's pH value are linearly correlated at a given temperature.

The advantages of a glass-membrane electrode are given as follows:

(1) It has a high selectivity to hydrogen ions and wide application that can be used to measure different hydrogen ion concentrations.

(2) It is not affected by oxidants and deoxidizers.

(3) The result is accurate and performs best when pH ranges from 1 to 9.

(4) It can be used not only for colored, turbid, and colloidal solutions but also for acid-base titration.

(5) The operation is simple, and the solution is not polluted. The response to the test is quick.

The disadvantages of a glass-membrane electrode are given as follows:

(1) It can be broken easily. A glass-membrane electrode is normally used for more than one year, so the capability would get worse for long-term use.

(2) The electric resistance is high.

(3) The measured pH value ranges from 1 to 10. It cannot be used for fluoride ions.

The hydrogen ion concentration is low for pH $>$ 9, or the sodium ion concentration is high. Thus, the potential of ion exchange from the electrode and solution interface is formed. However, the pH value will be much lower in a strong base solution, and the interference is more serious for high sodium ion concentration. This is called the **alkaline error.** It can be reduced by using lithium oxide instead of sodium oxide. This electrode is called the lithium glass-membrane electrode or high pH electrode. In this case, the maximum pH value can be 13.5, while the mechanical strength will be poor.

The error is also generated for pH$<$1, called the **acid error.** This is because H_2O transmits the hydrogen ion. As a result, the concentration of H_2O decreases, the hydrogen ion activity from the solution to the electrode surface gets lower, and the pH value is higher.

9.2.2 Measured by Potentiometry

In the analysis of pH potential, the indicated electrode is a glass-membrane electrode, while the saturated calomel electrode is the reference electrode. The electrochemical cell is given as:

Ag | AgCl, HCl | glass | sample solution | KCl (saturated), Hg_2Cl_2 | Hg

φ_{Mem} φ_{liq}

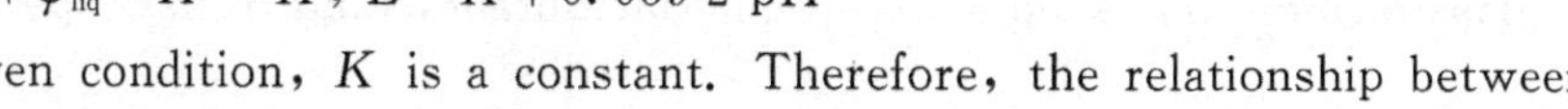

The liquid junction potential (φ_{liq}) is a constant for a given condition, formed between two solutions of different compositions or concentrations. The cell potential is given as:

$$E=\varphi_{Calomel}+\varphi_{liq}+\varphi_{glass}=\varphi_{Calomel}+\varphi_{liq}-(K'-0.0592\ pH) \quad \textbf{(Equation 9.10)}$$

If $\varphi_{Calomel}+\varphi_{liq}-K'=K$, $E=K+0.0592\ pH$

For a given condition, K is a constant. Therefore, the relationship between the cell potential and the pH value is linearly correlated. Because φ_{liq} and φ_{Asy} is unknown, the electrochemical cell potential cannot calculate the pH value. In experiments, a given pH solution is usually used as a standard solution for correction. A buffer solution of a given pH value is used to adjust the instrument before using the pH meter to measure the pH value. The purpose is to eliminate the effect of asymmetric potential.

The potential of the given pH solution is computed as:

$$\varphi_s = K + 0.0592\ \text{pH}_s$$

The potential of the sample solution is computed as follows:

$$\varphi_x = K + 0.0592\ \text{pH}_x$$

$$\varphi_x - \varphi_s = 0.0592(\text{pH}_x - \text{pH}_s)$$

$$\text{pH}_x = \frac{\varphi_x - \varphi_s}{0.0592} + \text{pH}_s$$

The pH meter can be divided into the potentiometer-type acidity meter and the direct reading type acidity meter.

9.3 Ion Selective Electrode

An **ion-selective electrode** (ISE) is an indicator electrode that can respond to individual types of anions or cations. There are several ion-selective electrodes, such as the sodium ion-selective electrode, sulfur ion-selective electrode, and calcium ion-selective electrode.

9.3.1 Principle of Ion Selective Electrode

An ion-selective electrode consists of a membrane, a supporter, an internal reference solution (e. g., texted ions), and an internal reference electrode (e. g., AgCl/Ag). Based on the membrane potential formed by the exchange of specific solution ions and membrane ions, the ion-selective electrode can be used to measure a specific type of ions. At a given condition, the Nernst equation of membrane electrode is given as:

For positive ion: $$\varphi_{\text{Mem}} = K + \frac{2.303RT}{nF}\lg \alpha_+$$

For negative ion: $$\varphi_{\text{Mem}} = K - \frac{2.303RT}{nF}\lg \alpha_-$$

The fundamental feature of an ion-selective electrode is the linear relationship between membrane potential and the logarithm of the activity of the tested ions for a given condition.

9.3.2 Type of Ion Selective Electrode

According to the materials used in the membrane, ion-selective electrodes can be divided into the glass-membrane, solid-state, liquid-membrane, and other special membrane electrodes.

9.3.2.1 pM glass electrode

The pM glass electrode is a glass membrane electrode selective for metal ions. The glass membrane of a sodium ion-selective electrode is composed of 11% Na_2O, 18% Al_2O_3, and 71% SiO_2. The glass membrane of a potassium ion-selective electrode consists of 27% Na_2O, 5% Al_2O_3, and 68% SiO_2. These electrodes can be used easily. However, they have high electric resistance and low selectivity.

9.3.2.2 Solid-state membrane electrode

A solid-state membrane electrode contains a sensing element that is a crystalline material or a homogeneous pressed pellet. The material composition can be divided into two types:

1. The single crystal membrane electrode

The fluoride ion-selective electrode is commonly used to measure fluoride. The electrode's sensing element is a lanthanum fluoride pellet (LaF_3). The Ag-AgCl electrode is the internal reference electrode in contact with 0.1 mol • L^{-1} NaCl and 0.1 mol • L^{-1} NaF solutions. The design is shown in Figure 9.2.

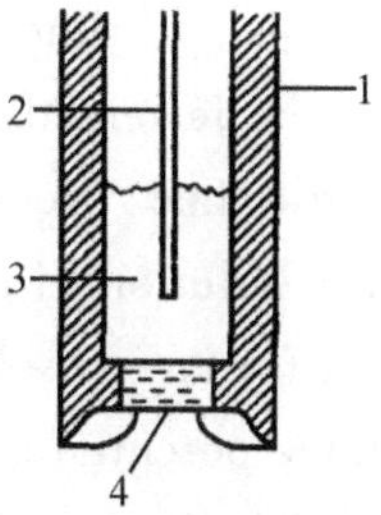

1. Plastic pipe 2. Ag-AgCl reference electrode 3. 0.1 mol • L^{-1} NaCl and 0.1 mol • L^{-1} NaF solution 4. LaF_3 crystal

Figure 9.2 Design of the LaF_3 crystal membrane electrode

2. Polycrystalline electrode

At high pressure, the membrane is made of powdered insoluble salts (e. g., AgCl, AgBr, AgI, Ag_2S). These powders can be multiple crystals. These electrodes include the Ag_2S membrane electrode (the membrane is made by Ag_2S precipitation. It is also the silver ion and sulfur ion-selective electrode.), AgX-Ag_2S membrane electrode (mainly used to measure halide ions), heavy metal sulfides-Ag_2S membrane electrode (the main material is Ag_2S and other metal sulfides). Some other electrodes include the Cu^{2+} selective electrode (CuS-Ag_2S), Cd^{2+} selective electrode (CdS-Ag_2S), and Pb^{2+} selective electrode (PbS-Ag_2S).

9.3.2.3 Liquid-membrane electrode

A liquid-membrane electrode consists of an inert porous substance infiltrated by an organic solvent containing the tested ion salts and chelates. The design is shown in Figure 9.3.

Ag-AgCl internal reference electrode is immersed in a 0.1 mol • L^{-1} Ca_2Cl solution. A non-aqueous, insoluble organic exchange solvent is the other solution used in the electrode.

The main factors affecting the ion-selective electrodes are listed below:

(1) **The Nernst equation** For most ion-selective electrodes, the relationship between the potential and concentration of tested ions follows the Nernst equation linearly, with the slope (S) of $2.303RT/nF$. At 25 ℃, $n = 1$, $S = 0.059$. However, some deviations from the Nernst equation may exist for a few

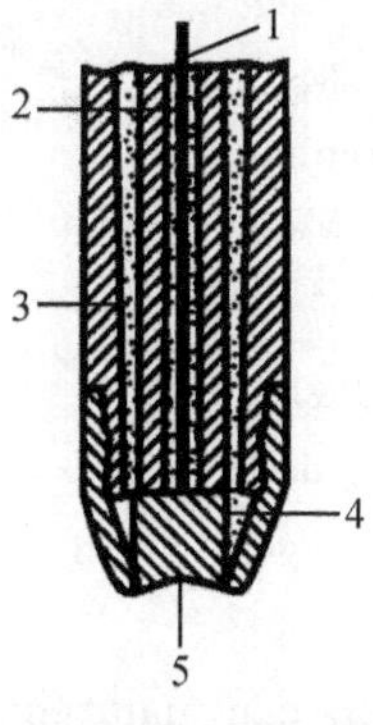

1. Ag-AgCl internal reference electrode 2. Internal solution 3. Organic liquid for ion exchange 4. Septal membrane 5. Hydrophobic membrane

Figure 9.3 Design of the liquid-membrane electrode

electrodes.

(2) **Selectivity** Although ion-selective electrodes only respond to the tested ions, some coexisting ions can affect the membrane potential. The **selectivity constant** (K_{AB}) is used to quantify the influence of the co-existing ions (B) on the main ions (A). For example, the selectivity constant of the hydrogen sensing electrode is 10^{-11}, indicating that the electrode response to H^+ is 10^{11} times more sensitive than to the Na^+ for the same concentration. The lower K_{AB} is, the fewer influences can be.

(3) **Response time** Response time is when an electrode reaches the stable potential after immersing in a solution. In 1976, IUPAC suggested the definition of response time as the time measured for an ion-selective electrode and a reference electrode immersed in a solution to attain a stable potential of 1 mV. Usually, the response time is very short and varies with concentration. The higher the tested concentration, the shorter the response time.

Other influencing factors include internal resistance, firmness of electrodes, and so on. Both gas-sensing electrodes and enzyme electrodes are common ion-selective electrodes. A **gas-sensing electrode** is a group of compound electrodes modified to analyze certain gases. In contrast, an **enzyme electrode** is a compound electrode manufactured using enzymes to convert analytes into products that can be measured by potentiometry.

9.3.3 Measurement of Ion Selective Electrode

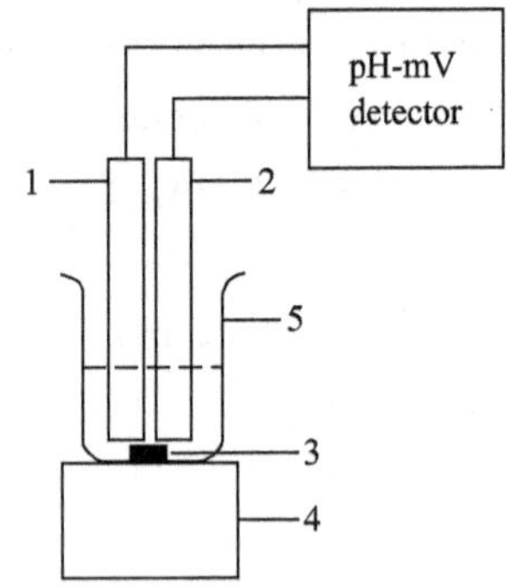

1. Ion selective electrode
2. Reference electrode
3. Agitator 4. Magnetic stirrer 5. Sample container

Figure 9.4 Design of the measuring device

The design of the measurement device is shown in Figure 9.4.

1. Standard curve method

At a given condition, the potentials of a series of standard solutions can be first measured. Then, a standard curve of E-lg α can be plotted on a coordinate paper. The relationship between the potential and the logarithm of the activity of the tested ion is a straight line. In the same condition, the potential of the sample is measured, and the sample concentration can be found according to the standard curve.

The relationship between the activity and concentration is shown below:

$$a = \gamma c \quad (\gamma\text{: activity constant})$$

For a given solution concentration, adding a total ionic strength adjustment buffer (i. e., TISAB, which is a high-concentration inert electrolyte solution) into the standard sample solutions can maintain the ionic strength and is a buffer and mask to the solution.

2. Standard addition method

For a tested ion with a concentration of c_x and volume of V_x, the measured potential is E_1. By accurately adding standard solution into the tested solution (the standard solution concentration c_s are more than $100c_x$; volume V_s is less than $100/V_x$.), the difference of concentration (Δc) is calculated as:

$$\Delta c = \frac{V_s c_s}{V_x} \geqslant c_x$$

The new potential is E_2. According to the Nernst equation,

$$E_1 = K + \frac{0.0592}{n} \lg c_x$$

$$E_2 = K + \frac{0.0592}{n} \lg(c_x + \Delta c)$$

Let $S = 0.0592/n$, then

$$E_2 - E_1 = \Delta E = S \lg \frac{c_x + \Delta c}{c_x} = S \lg\left(1 + \frac{\Delta c}{c_x}\right)$$

$$\frac{\Delta E}{S} = \lg\left(1 + \frac{\Delta c}{c_x}\right),\ 10^{\Delta E/S} = 1 + \frac{\Delta c}{c_x}$$

$$c_x = \frac{\Delta c}{10^{\Delta E/S} - 1}$$

The advantage of the standard addition method is that it does not need to plot the standard curve and can eliminate interference. However, the disadvantage is that it requires a long time and does not adapt to many samples.

9.3.4 Factors Affecting Measurement

1. Temperature

The electrochemical cell potential is given as:

$$E = K' \pm \frac{2.303RT}{nF} \lg \alpha_i$$

In this equation, the reference electrode potential and liquid junction potential are affected by temperature. Therefore, the slope of the straight line is related to temperature.

2. The accuracy of potential

The relationship between the potential difference (ΔE) and relative error of concentration $\Delta c/c$ is given as:

$$E = K' \pm \frac{RT}{nF} \ln c$$

$$\Delta E = \frac{RT}{nF} \times \frac{\Delta c}{c}$$

At 25 ℃ $$\Delta E = \frac{0.2568}{n} \times \frac{\Delta c}{c} \times 100$$

The relative error of concentration is given as:

$$\frac{\Delta c}{c}\times 100=\frac{n\Delta E}{0.256\ 8}\approx 4n\Delta E$$

If $n=\pm 1$ and $\Delta E=\pm 1$ mV, the relative error is $\pm 4\%$.

If $n=\pm 2$ and $\Delta E=\pm 1$ mV, the relative error is $\pm 8\%$.

The error of the direct potentiometric method is relatively high, which is higher for ions with a high charge number. Thus, the ion-selective electrode is usually adapted to the solution of low ion concentration. If the ion concentration is high, potentiometric titration should be used. The charge number of ions is (maximum) inversely proportional to the charge number (decrease).

3. Interfering ion

Co-existing ions would react on the membrane and form a new insoluble substance or directly react with the membrane. Adding masking agents can avoid the influence.

4. pH

The pH value is too high or too low, which will affect the measurement accuracy of the potential. Therefore, a buffer solution is often needed to adjust and maintain the hydrogen ion concentration of the solution during the measurement process.

5. Ion concentration

Usually, the detection range of ion-selective electrodes is 10^{-6} to 1 mol • L^{-1}. The detection limit is related to the membrane's activated materials. In addition, the co-existing ions and the concentration of hydrogen ions can also affect the ion-selective electrode detection range.

6. Response time

The response time of the ion-selective electrode is related to the activity of the tested ion, ionic strength of the medium, co-existing ions, thickness and surface smoothness of the membrane, and the rate of tested ions reaching a surface membrane.

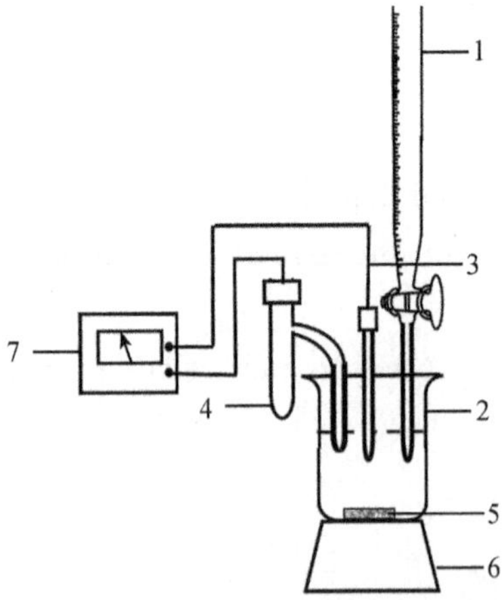

1. Burette 2. Titration cell 3. Indicator electrode 4. Reference electrode 5. Stirring magnet 6. Agitator 7. Potentiometer

Figure 9.5 Basic device of potentiometric titration

9.3.5 Application of Ion-selective Electrodes

Ion-selective electrodes can be applied to industrial production, environmental protection, medical science, geology, etc.

9.4 Potentiometric Titration

9.4.1 Device

Potentiometric titration utilizes a potential measurement in titration. The design of the measuring device is shown in Figure 9.5. A reference and an indicator electrode are immersed in a tested solution during titration. A certain number of titrants are added, and the potential is measured. A potentiometer or direct current voltmeter is used to measure the potential.

9.4.2 Determination of Titration Endpoint

During the titration procedure, the potential will be measured after adding titrants. A series of titrant volumes and potential values can be recorded until the potential exceeds the stoichiometric point. Near the stoichiometric point, the potential value varies frequently. Thus, it must be measured after every addition of 0.1-0.2 mL titrants. There are three methods to confirm the titration end point: the E-V method, $\Delta E/\Delta V$-V), and the second derivative method. The comparison between these methods is listed in Table 9.2.

Table 9.2 0.100 0 mol · L^{-1} $AgNO_3$ titrate NaCl solution

$V(AgNO_3)$/mL	E/mV	ΔE/mV	ΔV/mL	$\Delta E^2/\Delta V^2$ (mV^2/mL^2)
5.00	62			
		23	10.00	
15.00	85			
		22	5.00	
20.00	107			
		16	2.00	
22.00	123			
		15	1.00	
23.00	138			
		8	0.50	
23.50	146			
		15	0.30	
23.80	161			
		13	0.20	
24.00	174			
		9	0.10	
24.10	183			
		11	0.10	
24.20	194			2 800
		39	0.10	
24.30	233			4 400
		83	0.10	

Table 9.2 continued

$V(AgNO_3)$/mL	E/mV	ΔE/mV	ΔV/mL	$\Delta E^2/\Delta V^2$ (mV^2/mL^2)
24.40	316			−5 900
		24	0.10	
24.50	340			−1 300
		11	0.10	
24.60	351			
		7	0.10	
24.70	358			

1. The *E*-*V* curve method

On a plot of titrant volume (V) as abscissa potential (E) as ordinate, the inflection point of the E-V curve is the titration endpoint.

2. $\Delta E/\Delta V$-V method

Far from the stoichiometric point, the value of $\Delta E/\Delta V$ is small. Near the stoichiometric point, the value of $\Delta E/\Delta V$ increases rapidly. At the stoichiometric point, the $\Delta E/\Delta V$ value reaches the peak value. While over the stoichiometric point, the value of $\Delta E/\Delta V$ decreases gradually. According to the curve of $\Delta E/\Delta V$-V, the peak value is the titration endpoint.

3. The secondary derivative method

At the stoichiometric point, a dome shape can be observed in the magnified curve of $\Delta E/\Delta V$-V. By varying the volume, $\Delta E/\Delta V$ will keep a constant ($\Delta E^2/\Delta V^2=0$). On a plot with titrant volume (V) as abscissa and $\Delta E^2/\Delta V^2$ as ordinate, a curve of $\Delta E^2/\Delta V^2$-V is obtained. If $\Delta E^2/\Delta V^2=0$, the point is defined as the stoichiometric point.

The interpolation method is used to calculate the stoichiometric point directly. For example, if the titrant volume is 24.30, $\Delta E^2/\Delta V^2=$ 4 400. If titrant volume is 24.40, $\Delta E^2/\Delta V^2=$ −5 900. Using the interpolation method, we can calculate the stoichiometric volume point.

Titrant volume/mL	24.30	x	24.40
$\Delta E^2/\Delta V^2$	4 400	0	−5 900

$$(24.40-24.30)/(-5\,900-4\,400)=(x-24.30)/(0-4\,400)$$

$$(x-24.30)\times(-5\,900-4\,400)=(24.40-24.30)\times(0-4\,400)$$

$$(x-24.30)\times(-10\,300)=(-4\,400)\times(0.10)$$

$$x=\frac{(-4\,400)(0.1)}{-10\,300}+24.30=24.34(\text{mL})$$

At the stoichiometric point, the titrant volume is 24.34 mL.

The straight-line method is the method in which the titration curve is transformed into a straight line. This method, a straight line, can be drawn by several potentials near the stoichiometric point. These points can be far away from the stoichiometric point.

The straight-line method is based on all kinds of equilibriums (i. e., ionization equilibrium, mass balance, charge balance) in a solution. The equation is shown below:

$$V_e - V = K^{\ominus}_{a} \alpha_{H^+} V + \frac{V_0 + V}{c_B}(K^{\ominus}_{a} \alpha_{H^+} + 1)([H^+] - [OH^-]) \qquad \textbf{(Equation 9.9)}$$

V_e is the consumed titrant volume at the stoichiometric point. V_0 is the initial volume of the tested solution. V is the initial volume of the added solution. c_B is the titrant concentration. The $K^{\ominus}$ is the stability constant of a weak acid ($K^{\ominus} = 1/K^{\ominus}_a$). The α_{H^+} is the activity of hydrogen ions.

V and the pH value are obtained by experiments, which can be used to calculate the value of $V_e - V$. Then, a straight line of $V_e - V$ over V is plotted. The crossover point of the line at the V axis is the titrant volume at the stoichiometric point. For example, 0.099 61 mol · L^{-1} NaOH standard solution titrates 100 mL H_3BO_3 ($K^{\ominus}_{H_3BO_3} = 5.6 \times 10^{-10}$) solution. The data is given in Table 9.3, and the titration straight-line is shown in Figure 9.6.

Table 9.3 Data of NaOH solution titrates H_3BO_3 solution

V(NaOH)/mL	pH	$V_e - V$
0.00	6.71	4.038 6
0.50	8.28	3.546 2
…	…	…
4.50	9.89	0.698 3
5.00	10.17	0.292 7
5.50	10.56	−0.384 5
6.00	10.61	−0.864 9
6.50	11.15	−1.510 2
…	…	…
9.50	11.70	−5.508 2
10.00	11.74	−6.124 7

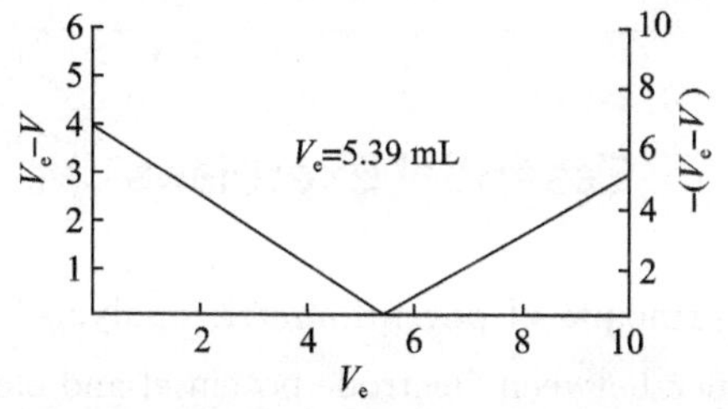

Figure 9.6 Titration straight-line

9.4.3 Applications of Potentiometric Titration

1. Acid-base titration

The pH glass-membrane electrode is an indicator electrode, while the calomel electrode is a reference electrode. It is a basic method in a non-aqueous solution for acid-base titration.

2. Redox titration

The platinum electrode is an indicator electrode, while the calomel electrode is a reference electrode. For example, $KMnO_4$ solution titrates I^-, NO_2^-, Fe^{2+}, V^{4+}, Sn^{2+}, $C_2O_4^{2-}$; $K_2Cr_2O_7$ solution titrates I^-, Fe^{2+}, Sn^{2+}.

3. Precipitation titration

The selection of indicator electrodes depends on the precipitation reaction. For example, $AgNO_3$ solution titrates halide ions. The silver electrode is used as an indicator electrode. The pH glass-membrane electrode is the reference, and a few HNO_3 were added to the solution. If a calomel electrode is used as the reference electrode, a KNO_3 salt bridge must separate the sample and calomel electrode. If the platinum electrode is an indicator electrode, a $K_4[Fe(CN)_6]$ solution can be used to titrate Pd^{2+}, Cd^{2+}, Zn^{2+}, and Ba^{2+}. AgX or Ag_2S ion-selective electrode can be used as an indicator electrode. $AgNO_3$ solution can be used to titrate Cl^-, Br^-, I^-, S^{2-}.

4. Coordination titration

The mercury electrode is an indicator electrode. EDTA solution can be used to titrate Cu^{2+}, Zn^{2+}, Ca^{2+}, Mg^{2+}, Al^{3+}. Fluoride ion-selective electrode as an indicator electrode can be used to titrate Al^{3+}. The calcium ion-selective electrode is an indicator electrode. EDTA can be used to titrate Ca^{2+}.

Key Glossary

Reference electrode: The reference electrode provides a fixed potential against which the potential of the indicator electrode can be measured. It is initially assigned the role of the anode used for potentiometry in an electrochemical cell.

Indicator electrode: The electrode in contact with the sample is the indicator electrode in an electrochemical cell for potentiometry. It gives a potential related to the activity and concentration of the analyte and is assigned the role of the cathode.

Essential Exercises

Review Questions

1. What is the basic principle of potentiometric analysis?
2. What is the difference between electrode potential and electrochemical cell potential?
3. Sum up the constitution, half-reaction, and potential of the indicator and reference electrodes.

4. What is the constitution and mechanism of the glass-membrane electrode?

5. What are the mechanisms and types of ion-selective electrodes?

6. What is potentiometric titration? What are the advantages and disadvantages compared with common titrations?

Cumulative Problems

1. An electrochemical cell is given as:

A glass-membrane electrode | H^+ ‖ SCE

The electrochemical cell potential is 0.200 V at 25 ℃ in a buffer solution of pH = 4.00. Two unknown concentration solutions replace the buffer solution, and the electrochemical cell potentials are 0.300 V and −0.020 V, respectively. What are the pH values of the unknown concentration solutions?

2. An electrochemical cell potential is 0.600 V. Therefore, the electrochemical cell is given as:

Pt, H_2 (1.01×10^5 Pa) | HA (0.100 mol • L^{-1}), NaA (0.100 mol • L^{-1}) ‖ SCE

Calculate the ionization constant of HA (liquid junction potential is ignored).

3. Measured the pH value by the following electrochemical cell:

A glass-membrane electrode | H^+ ‖ , a saturated calomel electrode

The electrochemical cell potential is 0.200 V at 25 ℃ in a buffer solution pH=4.00. Suppose the buffer solution is replaced with two solutions of unknown concentration. In that case, the electrochemical cell potentials are 0.212 V and 0.090 V. What are the pH values of the two solutions of unknown concentration?

4. At 25 ℃, the standard electrode potential of silver is 0.800 V. Calculate the potential in 0.1 mol • L^{-1} HCl solution.

5. A mobile carrier calcium electrode is used to measure the concentration of Ca^{2+}. The reference electrode is the cathode. The calcium electrode contacts with a 25.00 mL solution. The electrochemical cell potential is 0.469 5 V at 25 ℃. If 1.00 mL standard solution of $CaCl_2$ (the concentration is 5.45×10^{-2} mol • L^{-1}) is added into the solution, the electrochemical cell potential is 0.400 0 V. What is the concentration of the Ca^{2+} sample?

Chapter 10

Separation Techniques in Quantitative Analytical Chemistry

Key Points

- *Precipitation Separation Method*
- *Extraction Separation Method*
- *Ion Exchange Separation Method*
- *Chromatographic Separation Method*
- *Membrane Separation Technology*
- *Indicator Electrode and Reference Electrode*

10.1 Overview

In the quantitative chemical analysis process, direct measurement can be employed when dealing with samples of relatively simple composition, as discussed in these previous chapters. However, actual scenarios often present more complex situations, such as samples with intricate compositions, trace components needing measurement, or interference from other components. In such cases, appropriate separation methods are necessary to isolate and concentrate the components to be measured, ensuring they meet the requirements for accurate analysis.

The specific separation method used will vary depending on the nature of the components being tested. Various techniques, such as precipitation, solvent extraction, ion exchange, chromatography, membrane separation, etc., can be employed with the continuous development of separation technology and materials.

10.2 Precipitation Separation Method

The **precipitation separation** method, a cornerstone technique in analytical chemistry and chemical engineering, is predicated on the principle of solute solubility differential in a solvent under varying conditions. This method meticulously exploits the fact that the solubility of compounds in a solvent can significantly differ, allowing for the selective precipitation of a desired compound from a complex mixture. By carefully adjusting parameters such as temperature, pH, and the concentration of precipitating agents, a specific component can be induced to precipitate out of solution while others remain dissolved. The precipitated substance, often a solid, can then be isolated through filtration or centrifugation, providing a means to separate and purify compounds with high precision. This technique finds widespread application in the purification of proteins, the isolation of salts from aqueous solutions, and the recovery of precious metals from ores. Its efficacy and versatility make it an indispensable tool in pharmaceutical development, environmental analysis, and materials science, where the purity of substances is paramount. The precipitation separation method's fundamental yet profound approach underscores the importance of understanding chemical equilibria and solubility principles in strategically manipulating molecular interactions to advance scientific inquiry and technological progress.

Precipitation separation includes two methods: precipitation and co-precipitation.

10.2.1 Precipitation Method

The method that uses precipitation reactions for separation is the precipitation separation method, a classical separation technique. Its principle relies on the different solubilities of substances in a solvent to achieve separation. By adding an appropriate precipitant to the solution and controlling the reaction conditions, the component to be tested is precipitated, or the interfering components are removed by precipitation, thus separating the two. In qualitative analysis, the precipitation separation method is a primary means of separation. In quantitative analysis, since the solubility of any precipitate under certain conditions is always constant, the precipitation separation method can be used not only for separating major components but also for separating and enriching trace components using co-precipitation. Like the gravimetric method, if the precipitation separation method involves precipitating the component to be tested, it should be entirely precipitated without contamination from interfering components.

The precipitates are obtained using inorganic precipitants, except for a few crystalline precipitates like $BaSO_4$, which can quickly form larger particles. Most are amorphous or colloidal precipitates, such as the commonly separated hydroxides and sulfide precipitates. Small particles, large specific surface areas, loose structures, significant co-precipitation, and poor selectivity make the separation effect less than ideal. Organic precipitants, with their outstanding advantages, are widely used in precipitation separation methods.

The basic principle of the precipitation separation method is the precipitate solubility equilibrium. Discussions on the formation process of sediments, the conditions for precipitate formation, and the factors affecting co-precipitation have been covered in the "Gravimetric Method" and will not be reiterated here.

10.2.2 Co-precipitation

Co-precipitation separation is the method of separating, enriching, and purifying trace substances by precipitating them together with significant substances. The precipitant is divided into inorganic and organic co-precipitation methods based on the type.

1. Inorganic co-precipitation method

There are mainly two methods of separation using inorganic co-precipitants:

(1) Mixed crystal co-precipitation If the radius of the N ion (trace) to be separated in the solution is close to that of the M ion (major) in the co-precipitant M. NL and ML crystal structures are similar, so N can form mixed crystals NL-ML with ML and be carried down by ML.

(2) Adsorption co-precipitation In the method of surface adsorption co-precipitation, reagents that can form small particles of amorphous precipitates or colloidal precipitates, such as hydroxides or sulfides, are commonly used as co-precipitants. Due to the large specific surface area of small particle carriers, it is conducive to the adsorption of trace components to be separated.

2. Organic co-precipitation method

Organic co-precipitants have good selectivity, forming precipitates with lower solubility, more complete precipitation, and purer quality. Additionally, organic co-precipitants are easy to decompose and volatilize upon calcination, eliminating the interference of the precipitant on the substance to be tested while enriching the element to be tested, thus receiving broader application and development. Separation and enrichment using organic co-precipitants can be broadly divided into three types:

(1) Co-precipitation using colloidal aggregation, such as animal glue, tannin, and cinchonine, with co-precipitated components, including elements like tungsten, niobium, and silicon.

(2) Co-precipitation forming ion-association complexes, such as methyl violet with InI_4^-. Commonly used in acidic solutions are organic compounds such as methyl violet, malachite green, magenta, and methylene blue, which exist as cations, and ligands like Cl, Br, I, and SCN acting as co-precipitants for metal ions, with co-precipitated metal ions including Zn, Cd, Hg, Bi, Au, and Sb.

(3) Co-precipitation using inert co-precipitants, akin to the role of organic extractants in solvent extraction, can be understood as using a "solid extractant" for co-precipitation. The commonality between ion-association complexes and inert co-precipitants lies in first converting inorganic ions into hydrophobic compounds, allowing them to enter and be co-precipitated out by a structurally similar carrier, based on the principle of "like dissolves like".

10.3 Extraction Separation Method

The method known as solvent **extraction separation**, or simply extraction separation, utilizes the differing solubility of substances in water to separate them. Typically, the property of a substance being soluble in water but insoluble in non-polar organic solvents is termed hydrophilic, while the opposite property is termed hydrophobic. By shaking an organic solvent, which is immiscible with water, together with an aqueous solution, the two phases are brought into full contact, allowing substances with different affinities for water to redistribute themselves between the aqueous and organic phases. Hydrophilic substances remain in the aqueous phase, while hydrophobic substances enter the organic phase. Separating the two phases also separates the hydrophilic and hydrophobic substances. A substance moving from the aqueous phase to the organic phase is called extraction, while the opposite process is called back-extraction. Since the process does not involve chemical changes, extraction separation is considered a physical process.

10. 3. 1 Fundamentals

Utilizing the difference in solubility or partition coefficient of compounds in two immiscible (or slightly soluble) solvents allows for the transfer of compounds from one solvent to another. Most of the compound can be extracted through repeated extractions. The distribution law serves as the primary theoretical basis for the extraction method, as substances have different solubilities in different solvents. Moreover, when a certain soluble substance is added to two immiscible solvents, it can dissolve in both solvents. Experiments have shown that, at a constant temperature, when the compound does not undergo decomposition, electrolysis, association, or solvation with these two solvents, the ratio of this compound in the two liquid layers is a constant value. This is true regardless of the substance added and is considered a physical change.

The greater the solubility or partition coefficient of the solute in the two phases, the higher the extraction efficiency. A single extraction is usually insufficient to completely extract the desired solute from a solution; it must be repeated several times.

10. 3. 2 Applications

The extraction separation method has the advantage of effective separation, achieving high recovery rates through repeated extractions. The operation of the extraction separation method is straightforward and conducive to automation, with a wide range of applicability. It is suitable not only for the separation of components in substantial quantities but also for the separation and enrichment of trace components, not only for the separation of small samples in laboratories but also for the separation and purification of large amounts of substances in industrial production. If the extracted component is a colored compound, the organic phase can be directly used for spectrophotometric determination, known as extraction spectrophotometry, which offers high selectivity and sensitivity. Furthermore, the solvent extraction separation method is an indispensable pretreatment step for atomic absorption spectroscopy, chromatographic methods, and electrochemical analysis techniques.

10. 3. 3 Precautions

Organic compounds generally have a higher solubility in organic solvents than in water. Extraction separation follows the principle of "like dissolves like". For example, butanol with higher polarity is suitable for extracting saponin compounds; ethyl acetate is often used to extract flavonoid components; trichloromethane's hydrogen atoms can easily form hydrogen bonds with nitrogen atoms, making it suitable for extracting alkaloid components; volatile oils and other non-polar components are appropriately extracted with solvents such as petroleum ether and ether.

During extraction, the addition of a certain amount of electrolyte (such as sodium chloride) to the aqueous solution can utilize the "salting-out effect" to reduce the solubility of organic substances and the extraction solvent in the aqueous solution, often enhancing the extraction effect.

To extract weakly acidic components, the pH of the aqueous solution should be below the weak acid's pK_a by 1-2 units, while for extracting weakly basic components, the pH should be above the weak acid's pK_a by 1-2 units, ensuring the full ionization of weakly acidic and basic components.

Extraction is commonly conducted in a separatory funnel, where care should be taken to release gas and prevent emulsification; shaking should not be too vigorous, especially during the first extraction. If emulsification occurs, it can be eliminated by adding a small amount of ethanol or organic solvent or centrifugation. Multiple extractions should follow the principle of small amounts many times, generally requiring 3-4 extractions to completely extract the component of interest.

Because the solvents used in extraction separation are often flammable, volatile, and somewhat toxic, and most extraction solvents are expensive, their application is somewhat limited. Furthermore, based on traditional liquid-liquid extraction technology, liquid-phase microextraction technology, which uses less solvent, is environmentally friendly, and is widely applied in analyzing biological samples, environmental samples, and food, has been developed. However, the reproducibility of results needs improvement.

10. 4 Ion Exchange Separation Method

Ion exchange separation is a technique that utilizes the exchange of anions and cations to achieve separation. This process involves using an insoluble polymer compound whose molecules contain dissociable groups capable of exchanging with other cations or anions in the aqueous solution. By employing this method, the desired separation can be accomplished. Ion exchange separation has extensive applications in separating inorganic ions, organic compounds, and biochemical substances.

10.4.1 Structure and Properties of Ion Exchangers

10.4.1.1 Structure of ion exchangers

Ion exchangers have two main components: an inert skeleton and an exchangeable group. The inert skeleton carries the exchange groups and undergoes no exchange reactions during separation. However, its structure and properties, such as particle size, distribution, internal pore size, and specific surface area, greatly influence the separation performance. Framework materials can be categorized into two types: inorganic and organic. Inorganic ion exchangers, such as puff stone ($Na_2Al_2Si_4O_{12} \cdot nH_2O$), clay, and certain metal oxides, have inorganic materials as the skeleton. On the other hand, organic ion exchangers, specifically ion exchange resins, have organic materials as the skeleton. This section will focus on ion exchange resins.

Ion exchange resin is a synthetic polymer compound that typically exists in the form of spheres or amorphous granules. The polymer comprises a crosslinked network structure formed by styrene and divinylbenzene (DVB) (Figure 10.1).

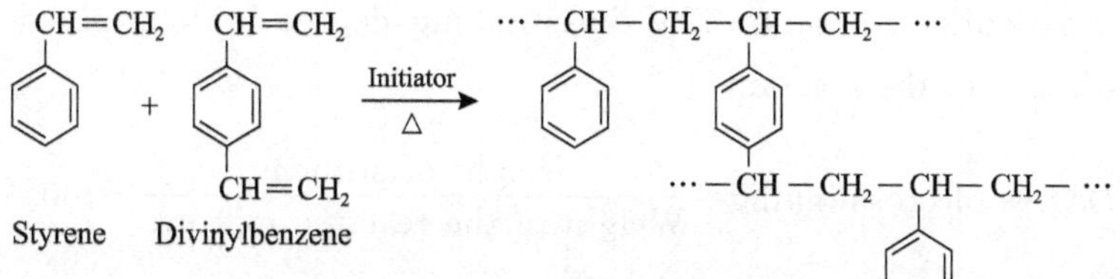

Figure 10.1 The framework of ion-exchange resin

Subsequently, the resin skeleton undergoes sulfonation or amination reactions to obtain ion exchange resins with different groups (Figure 10.2).

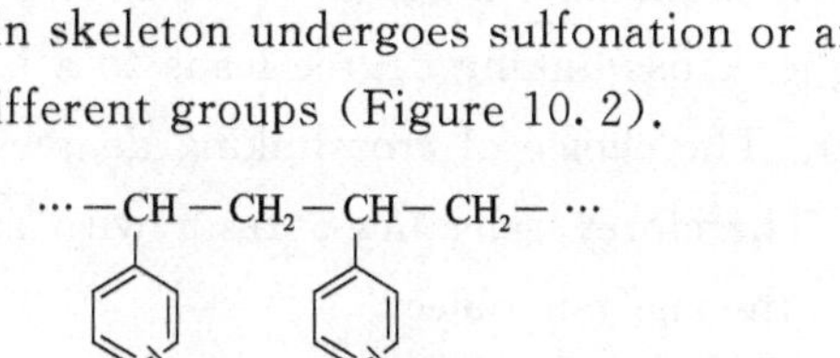

Figure 10.2 The structure of polystyrene sulfonic acid type ion exchange resin

Based on the exchange groups, ion exchange resins can be classified as cation and anion exchange resins. Additionally, different types of resins can be further categorized as strong, medium, weak, etc., based on their dissociation properties (Table 10.1).

Table 10.1 Basic classification of ion exchange resins

Ion exchange resin	Type	Exchangeable groups are commonly used
Cation exchange resin	Strong acid cation exchange resin Medium acid cation exchange resin Weak acid cation exchange resin	—SO_3H (sulfonic acid) —$PO(OH)_2$ (phosphate group) —COOH (carboxylate group)
Anion exchange resin	Strong base anion exchange resin Medium base anion exchange resin Weak base anion exchange resin	—$N^+(CH_3)_3OH^-$ (quaternary amine) —$N^+(CH_3)_2C_2H$ OH OH^- (quaternary amine) —NH_2 (amine, polyamine)

In addition, it contains active groups capable of forming chelates with specific metal ions, such as —$N(CH_2COOH)_2$. This resin can selectively exchange certain metal ions under specific conditions, making it highly significant in chemical separation.

10.4.1.2 Properties of ion exchange resin

1. Degree of crosslinking

The styrene in the core of the ion exchange resin is crosslinked with stilbene to form a macromolecular network structure. The cross-linking degree is determined by the weight percentage of stilbene in the reactant.

$$\text{Degree of crosslinking} = \frac{\text{Weight of stilbene}}{\text{Weight of the reaction mixture}} \times 100\%$$

The degree of crosslinking directly affects the tightness and pore size of the resin skeleton's network structure, affecting the speed of exchange reactions and selectivity. More cross-linking leads to a smaller mesh size, a tighter texture, and a reduced tendency to swell in water. Conversely, a higher cross-linking degree leads to a larger mesh size and greater propensity to swell in water. The choice of crosslinking degree depends on the size of the molecules being separated. Therefore, selecting a resin with an appropriate cross-linking degree is important based on the analysis object.

2. Exchange capacity

The exchange capacity of ion exchange resin is determined by the number of ion exchange groups expressed in exchange equivalents. The exchange capacity represents the milliequivalents of exchangeable ions per gram of dry resin. In 1972, the International Union of Pure and Applied Chemistry (IUPAC) proposed the following symbols for expressing exchange equivalents:

Q_A—analyzed weight exchange capacity. It indicates how many mmol of exchangeable H^+ or OH^- is contained in 1 g of dry resin under given conditions (usually stated).

Q_B—penetrating capacity. It represents the actual exchange capacity of a known amount of ion exchange resin packed in the column. In specific conditions, a solution containing an ion is passed through the exchange column until the concentration of this ion in the eluent reaches a specified concentration. It is expressed as the number of millimoles (mmol) of

H^+ or OH^- ions that flow through 1 g of dry resin.

The composition of the skeleton significantly affects the exchange capacity. For instance, when the crosslinking degree of Dowex 1Cl-type strong base anion exchange resin changes from 2 to 12, the weight exchange capacity decreases from 4.3 mmol • g^{-1} to 2.7 mmol • g^{-1} of dry resin. Exchange capacity is a significant indicator for resins when separating larger quantities of substances.

3. Swelling property

Ion exchange resin possesses polar active groups and exhibits hydrophilic properties. When the resin encounters water in its dry state, it absorbs it and undergoes swelling. The following factors generally influence the extent of resin swelling:

(1) The ease of dissociation of exchangeable groups (such as strong acidity or alkalinity) results in higher hydration and greater swelling.

(2) The degree of crosslinking is also a determining factor—resins with higher crosslinking exhibit greater swelling. Resins with smaller mesh sizes have a denser structure and are less prone to swelling in water, whereas those with larger mesh sizes and higher crosslinking are more likely to swell.

(3) The nature of the external solution also affects swelling. Resins exhibit higher swelling in solutions with lower electrolyte concentrations.

10.4.2 Application of Ion Exchange Separation

Ion exchange technology is crucial in various domains, including the chemical industry and biomedicine. It finds applications in industrial wastewater purification, metal smelting, separation and purification of active substances, and virus removal.

10.4.2.1 Application of ion exchange resin in wastewater treatment for nitrogen removal

Ion exchange is a commonly used method for treating nitrate wastewater. It effectively removes pollutants through two main mechanisms. Firstly, charged groups are exchanged with the functional groups of ion exchange resin through electrostatic force. Secondly, uncharged groups are exchanged through van der Waals or hydrophobic interaction forces, which adsorb to the resin surface. The goal of pollutant removal can be achieved by utilizing the adsorption and exchange effect of ion exchange resin on pollutant ions in water. For instance, the strong alkaline anion exchange resin D890 can reduce the nitrate nitrogen in incoming water from approximately 10 mg • L^{-1} to less than 1 mg • L^{-1} at a flow rate of 10 times the resin volume of the incoming water per hour (i.e., 10 BV • h^{-1}).

Moreover, the regeneration process is economical and effective, requiring only a sodium chloride mass fraction of 4%. When using D890 resin, the removal rates of total nitrogen and nitrate nitrogen in the secondary biochemical effluent of urban sewage treatment plants can reach 88.6% and 89.6%, respectively, with the total nitrogen in the effluent being lower than the limit value of Class Ⅳ defined by the *Environmental Quality Standard for Surface Water* (GB 3838—2002). Additionally, D890 resin has a positive charge, enabling it to adsorb most negatively charged microorganisms. After adsorption, the relative abundance of Proteobacteria increased from 35.41% to 91.40%, while the number of other communities decreased significantly. The operable classification units before and after adsorption were 659 and 247, respectively, reducing 62.5%. Therefore, resin adsorption significantly removes bacteria in the water, which is beneficial for subsequent effluent disinfection.

10.4.2.2 Isolation of active parts of natural products

Ion exchange separation can be utilized to separate various substances such as amino acids, peptides, alkaloids, organic acids, phenols, and more. For instance, aqueous solutions are commonly used to extract traditional Chinese medicine. The resulting water extract can then undergo a two-step process involving a strongly acidic resin (e.g., sulfonic acid type) followed by a strongly alkaline resin (e.g., quaternary amine type) to achieve separate elution. This elution process allows for dividing the extracted components into acidic, neutral, and alkaline parts, which can be further evaluated through animal or clinical trials (Figure 10.3).

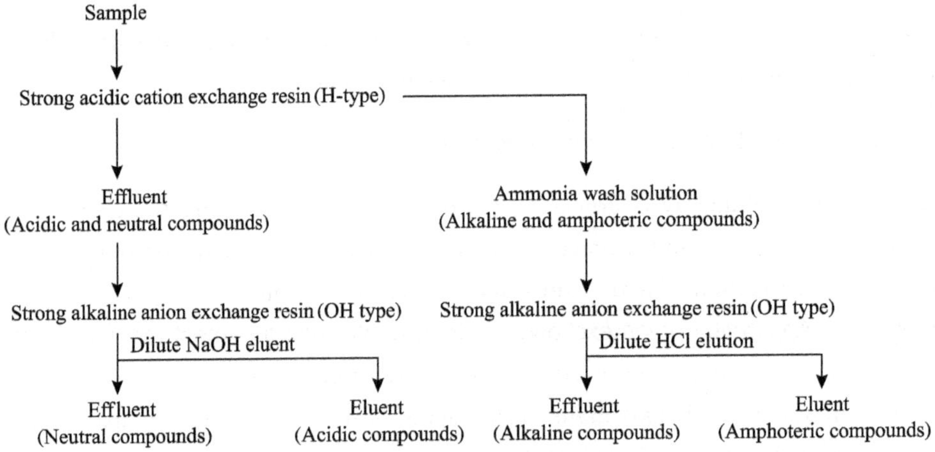

Figure 10.3 Mode diagram of ion exchange resin method for separating substances

The isolation of alkaloid components in traditional Chinese medicine involves using strong acidic resin to extract alkaloids from Chinese medicinal water extracts or dilute ethanol extracts. The alkaloid parts can be obtained by elution with ammonia water or ammoniacal ethanol.

10.4.2.3 Isolation and purification of proteases

In the past, packed bed column chromatography was commonly used for separating and purifying proteins. However, it had limitations due to the adsorbent used, resulting in low productivity of the desired protein. Recently, membrane chromatography has emerged as a successful solution to overcome these challenges. Functional membranes with ion exchange capabilities can be created. These functional membranes effectively adsorb the required proteins, enabling efficient separation and purification. Some researchers have utilized electrospinning technology to develop a polyacrylonitrile (PAN) nanofiber membrane with a three-layer structure. This membrane can be easily modified through chemical grafting, creating a multifunctional nanofiber membrane. Hydrolyzing PAN produces a weak ion exchange nanofiber membrane, which can be further coupled with aminomethanesulfonic acid to obtain a strong nanofiber membrane. The study investigated the purification efficiency of lysozyme using these two different ion exchange nanofiber membranes. The results demonstrated that the weak ion exchange membrane achieved a lysozyme recovery rate of approximately 93% and increased purity by about 29 times. Similarly, the strong ion exchange membrane achieved a lysozyme recovery rate of around 90% and increased purity by about 36 times. These findings indicate that both ion exchange membranes can be adsorbents in lysozyme purification.

10.4.2.4 Isolation and purification of the virus

Viral infections have always been a significant threat to the health of humans and livestock worldwide. To combat these infections, it is crucial to employ efficient manufacturing procedures that ensure the viral safety of biopharmaceutical products, such as blood products. Anion exchange chromatography (AIEX) is the most commonly used chromatography technique for virus purification. This technique adsorbs negatively charged viral surface proteins to positively charged ligands, allowing all positively charged molecules to pass through the packed column and exit the process. The adsorbed material is subsequently removed from the matrix, and the adsorbed material is eluted from the matrix by increasing salt concentrations.

10.5 Chromatographic Separation Method

10.5.1 Fundamental Concepts

Chromatographic separation, commonly called **chromatography**, is also known as layer analysis and color layer method. Chromatography is a multi-stage separation technique based on the minute differences in the distribution coefficients of the substances being separated in two phases (stationary and mobile phases). As the two phases undergo relative movement, the analyte undergoes repeated distribution, causing the initially subtle distribution differences to amplify further and separate individual components. This method demonstrates high separation efficiency and can effectively separate substances with very similar properties. Chromatographic separation methods encompass various types, and this section will exclusively introduce the widely used ones based on chemical principles, including extraction chromatography (reverse-phase distribution chromatography), thin-layer chromatography, paper chromatography, liquid chromatography, and gas chromatography. The subsequent volume will comprehensively discuss other chromatographic separation methods (instrumental analysis).

10.5.2 Extraction Chromatography

10.5.2.1 Principles of Method

Extraction chromatography, a liquid-phase distribution chromatography combining solvent extraction with chromatographic separation techniques, is typically conducted in a column format known as column chromatography. In this method, the stationary phase consists of organic extractants either impregnated or retained on a porous, hydrophobic, inert carrier. In contrast, the mobile phase comprises an aqueous solution containing suitable inorganic compounds. This essentially involves the aqueous solution's reverse extraction of analytes from the organic phase, hence reverse-phase distribution chromatography. The test solution containing the components to be separated is placed on the upper layer of the chromatographic column. As the mobile phase flows, the separated components gradually move downward. Simultaneously, they undergo repeated extraction and reverse-extraction between the two phases. The final evaluation of separation quality is determined based on the elution curves of each separated component.

10.5.2.2 Key Factors Influencing Separation

The primary factors influencing separation include the stationary, carrier, and mobile phases. The stationary phase consists of organic extractants and organic molecules chemically bonded to the silicon hydroxyl groups on the silica surface, which form another type of chemically bonded stationary phase. With a wide variety of organic extractants available, the choice depends on the nature of the components to be separated. The extractant, serving as the stationary phase, should firmly adsorb to the carrier and remain insoluble in the mobile phase.

The carrier must be inert, porous, uniformly distributed in pore size, and have a large surface area and excellent physical and chemical stability. It should not swell or adsorb ions from aqueous solutions, be heat-resistant, and resist acid and alkali corrosion. Carrier materials include diatomaceous earth, silica gel, polytetrafluoroethylene, polyethylene-co-acrylic ester copolymers, foam plastics, activated carbon fibers, etc. The carrier is immersed in an organic solvent solution of the extractant, followed by drying after the organic solvent evaporates, before column packing.

Extraction chromatography commonly employs inorganic acid solutions as the mobile phase, and the type and concentration of the inorganic acid significantly affect the separation factors. Occasionally, the addition of inorganic salts is employed to alter separation outcomes. Nitric acid, hydrochloric acid, sulfuric acid, and mixed solutions of these acids with inorganic salts are frequently used. As the mobile phase is an aqueous solution, separation efficiency can be enhanced by adjusting the acidity of the mobile phase and incorporating ligands. Specifically, using certain ligand-containing mobile phases facilitates the reverse extraction of some components, contributing to the separation. Extraction chromatography is often utilized in inorganic analysis.

10.5.2.3 Operational Procedures

The operational procedures are as follows:

(1) Preparation of stationary phase and equilibration The stationary phase is prepared by impregnation. The organic extractant is dissolved in a volatile solvent to form a suitable concentration solution. The carrier with the appropriate particle size is added, and the mixture is stirred or shaken for a specified duration. After solvent evaporation, the stationary phase is obtained. Column packing requires a uniform, compact filling without air bubbles. Equilibration is performed by passing 5-10 times the carrier volume of the mobile phase through the chromatographic column.

(2) Sample Introduction The sample solution is adjusted to the optimal conditions for extraction and introduced onto the column. The flow rate through the chromatographic column is controlled.

(3) Washing An aqueous solution with a flow rate like the sample solution is used to wash the column bed, removing unextracted components.

(4) Elution and separation With controlled temperature and an appropriate flow rate, the separated components are eluted using the mobile phase. The optimal composition of the mobile phase is chosen based on the nature of the separated components, ensuring complete separation. Stepwise elution methods are often employed. For example, using P507-carbon fiber as the stationary phase for separating rare earth elements and washing with different concentrations of hydrochloric acid results in the basic separation of 15 rare earth elements. Using a column with n-octylamine-cellulose, elution with 10 mol · L^{-1} HCl, 6 mol · L^{-1} HCl, and 0.05 mol · L^{-1} HNO_3 separately achieves complete separation of Th^{4+}, Zr(Ⅳ), and UO_2^{2+} (Figure 10.4).

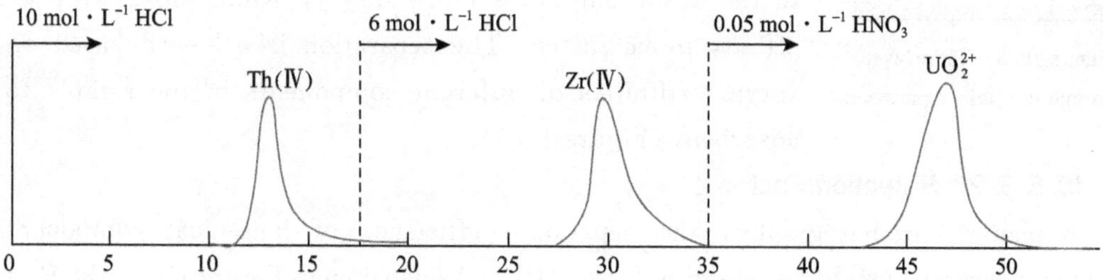

Figure 10.4 Chromatographic separation of Th^{4+}, Zr(Ⅳ), and UO_2^{2+} mixture

(5) Regeneration of the chromatographic column There are two methods for regeneration. The first method involves removing the carrier with the attached stationary phase, washing away the extractant with an organic solvent, heating to remove the residual organic solvent, and reimpregnating the carrier with the stationary phase. The second method involves adding a solution of the extractant in an organic solvent at the top of the column, slowly passing it through the chromatographic column, immersing the carrier attached to the stationary phase, and then washing away the excess organic phase with water.

10. 5. 3 Thin-Layer Chromatography and Paper Chromatography

Thin-layer chromatography (**TLC**) and **paper chromatography** fall under the planar chromatography category, as the shape of the stationary phase is planar. Thin-layer chromatography involves a liquid-phase chromatographic method with an adsorbent thin layer fixed on a planar support (such as a glass plate). On the other hand, paper chromatography uses filter paper as the carrier, with the stationary phase being the water absorbed on the filter paper. Both methods are liquid-phase chromatography, with paper chromatography becoming less commonly used. This section will primarily focus on thin-layer chromatography.

10. 5. 3. 1 Method Principles

In thin-layer chromatography, the stationary phase consists of an adsorbent (such as silica gel, activated alumina, cellulose, etc.), typically uniformly coated on a glass plate to form a thin layer. The sample is then spotted at one end of the thin layer, a certain distance away from the edge, allowed to dry, and the thin-layer plate with the sample is immersed into the mobile phase, acting as the eluent (avoiding immersion of the sample spot). The mobile phase is in a covered chromatographic tank with a saturated vapor pressure. Due to the capillary action of the stationary phase adsorbent, the mobile phase ascends along the thin layer of the stationary phase. When it encounters the sample spot, it dissolves in the mobile phase, and a series of adsorption-desorption processes occur between the mobile and stationary phases. Substances with stronger affinity to the adsorbent move more slowly, while those with weaker affinity move faster. The separation is achieved based on the varying affinities of different components in the sample to the adsorbent (Figure 10. 5).

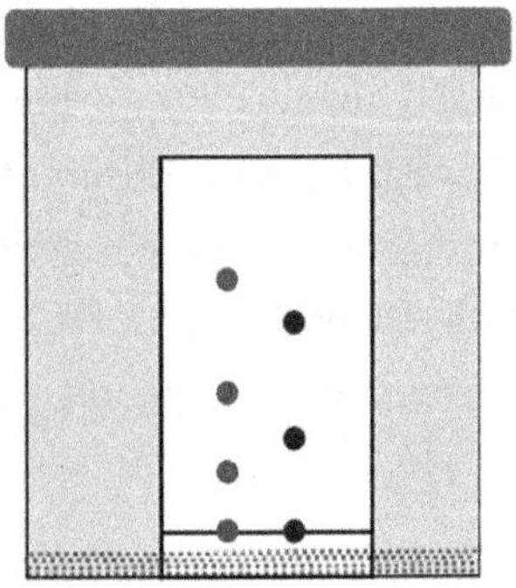

Figure 10. 5 Thin-layer chromatographic separation

10. 5. 3. 2 Retention Factor

In planar chromatography, the separation efficiency of individual components is commonly assessed using the retention factor (R_f). As depicted in Figure 10. 6, the R_f value is calculated by the formula:

$$R_f = \frac{a}{b}$$

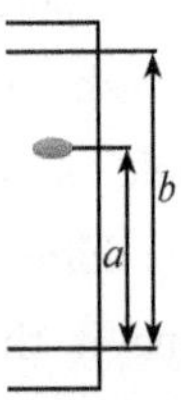

Figure 10. 6 Measurement of migration distance

where a is the distance from the center of the spot to the origin (in cm), and b is the distance from the solvent front to the origin (in cm). The maximum R_f value is equal to 1, indicating that the component moves with the solvent front, signifying a huge distribution ratio D. Conversely, the minimum R_f value is equal to 0, meaning that the component remains stationary at the origin, implying a minimal distribution ratio D. The more significant the difference in R_f values between two components, the better the separation efficiency.

10.5.3.3 Stationary Phase

In thin-layer chromatography, the stationary phase includes adsorbents such as silica gel, activated alumina, cellulose, and polyamide coated on inert thin layers for adsorption distribution separation. The stationary phase should possess a certain surface area, mechanical strength, and stability. It should remain insoluble in the mobile phase and exhibit reversible adsorption capabilities. The most used adsorbents for TLC are silica gel and activated alumina. Activated alumina comes in three types: alkaline, neutral, and acidic, suitable for separating alkaline, neutral, and acidic substances, respectively. Alumina tends to have high water content, indicating low activity. The particle size of activated alumina for TLC is generally around 200 mesh. Silica gel is divided into Silica Gel G and Silica Gel H, with the former containing gypsum as a binder. The activation degree of silica gel thin layers with binders is directly proportional to their adsorption capacity. The order of adsorption affinity for some compounds is as follows:

Saturated hydrocarbons < Unsaturated hydrocarbons < Ether < Ester < Aldehyde < Ketone < Amine < Hydroxyl compounds < Acids and bases

10.5.3.4 Mobile Phase

In normal-phase thin-layer chromatography, the mobile phase typically consists of organic solvents with slight amounts of acid or base. In reversed-phase chromatography, inorganic acid aqueous solutions are commonly used as the mobile phase. The choice of eluent depends on the nature of the sample and the separation mechanism. For adsorption-type thin-layer chromatography, the polarity of the mobile phase is crucial, and its polarity is directly proportional to its elution ability. The polarity order of some major pure solvents is as follows:

Petroleum ether < Cyclohexane < Carbon tetrachloride < Benzene < Toluene < Dichloromethane < Chloroform < Ether < Ethyl acetate < Acetone < Propanol < Ethanol < Methanol < Pyridine < Acid

The mobile phase can be a single solvent or a mixture of solvents to adjust the polarity of the mobile phase. Generally, one should choose an adsorbent with low activity and a mobile phase with high polarity for polar substances. An adsorbent with high activity and a low or non-polar mobile phase should be selected for non-polar substances.

10.5.3.5 Quantitative Measurement Methods and Applications

After selecting stationary and mobile phases, spotting, development, and separation, the thin-layer plate is removed, and the solvent front is marked with a pencil. If the separated components are colorless, coloration is performed, and the colored spots of the separated components are marked. Semi-quantitative analysis is generally carried out by measuring the area of the spots, comparing color intensity, and comparing with standard substances. Additionally, the colored spots can be scraped off, dissolved in an appropriate solvent, filtered, and made up to volume, and then the content is determined using suitable analytical methods. Thin-layer scanning spectrophotometers or fluorescence spectrophotometers can be used directly to measure the absorbance and fluorescence intensity of spots, determining the content of the analyte.

Thin-layer chromatography is widely used for its simplicity and is particularly suitable for analyzing and detecting organic components. For example, it can be employed to separate and detect vitamin B_1, vitamin B_2, and vitamin B_6 in plants, using a stationary phase of silica gel G impregnated with 0.5 % CMC (carboxymethyl cellulose) in a water solution (1 g : 3 mL) and a mobile phase (eluent) of chloroform-ethanol-acetone-ammonia solution (20 : 20 : 20 : 10, volume vatio), and an extraction solution of water-saturated n-butanol-6% TritonX-100 ethanol solution (23 : 2, volume ratio), with quantification using a thin-layer dual-wavelength scanning spectrophotometer.

10.5.4 High-Performance Liquid Chromatography

10.5.4.1 Principles of Method

High-performance liquid chromatography (HPLC) involves separating components dissolved in the mobile phase as they interact with the stationary phase. This interaction, which can be adsorption, distribution, exclusion, or affinity, varies in strength and size, resulting in different retention times within the stationary phase. HPLC has a broad application range, making it suitable for various samples without limitations on volatility and thermal stability. HPLC allows for the analysis and determination of a wide range of compounds, encompassing high-boiling, polar, ionic, and substances with large molecular sizes, overcoming the constraints associated with gas chromatography.

10.5.4.2 Structure and Components of HPLC Instrumentation

HPLC instruments typically consist of an output pump, injection device, chromatographic column, gradient flush device, detector, and data processing and microcomputer control unit (Figure 10.7). The output pump continuously delivers the flush solution to the column system at high pressure, facilitating the separation process of mixed samples in the chromatography. Common injection methods include diaphragm, valve, and automatic sampler. The chromatographic column serves to separate the components of the mixture. Gradient flushing, also known as solvent programming, enhances the resolution of complex samples and shortens the analysis cycle by continuously adjusting the composition of the flushing solution. The detector displays or converts the separated components from the chromatographic column into corresponding electrical signals. Commonly used detector types include UV absorption detectors, fluorescence detectors, electrochemical detectors, and refractive index detectors, with UV absorption detectors being the most widely used.

10.5.4.3 Factors Influencing Separation

The main factors influencing separation in gas chromatography (GC) include the type and

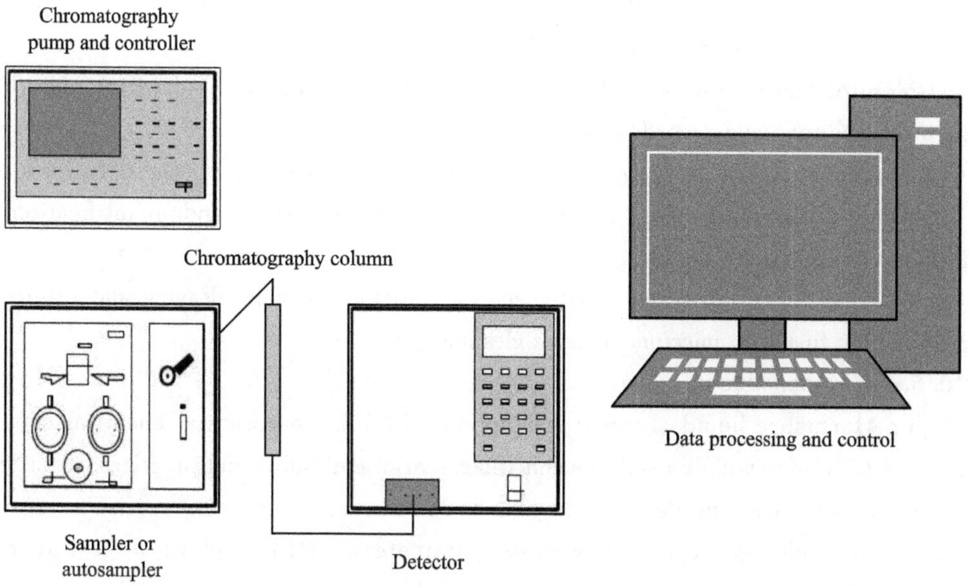

Figure 10. 7 High-performance liquid chromatography (HPLC) chromatogram

properties of the stationary phase, solvent system, flow rate, and sample properties. The stationary phase's type, particle size, and surface properties directly affect separation efficiency. Different types of stationary phases have different chemical properties and affinities, affecting the selectivity and rate of separation products. The solvents' composition, polarity, and flowability in the solvent system directly influence how sample molecules interact with the stationary phase, thereby affecting separation efficiency. Too-fast and too-slow flow rates can impact separation because they alter the time compounds interact with the stationary phase, affecting the degree of separation of various compounds. Sample properties, such as molecular size, polarity, and chemical activity, influence their distribution in the chromatographic column and separation efficiency. Different sample characteristics require the selection of appropriate separation conditions.

10. 5. 4. 4 Operating Procedure

The operating procedure is as follows:

(1) Prepare equipment and samples Ensure the chromatographic column and system are in good condition, fill the chromatographic column, and connect the detector. Dissolve the sample in an appropriate solvent and filter to remove impurities.

(2) Prepare the mobile phase Configure different solvents and mobile phase ratios as needed. Ensure the solvent is of high quality with no bubbles or impurities.

(3) Set chromatograph parameters Start the chromatograph and set parameters such as flow rate, detector sensitivity, and temperature.

(4) Sample injection Introduce the sample into the chromatographic column using an appropriate volume injector. Automatic and manual injectors are the two common methods, with automatic injectors being more commonly used.

(5) Start separation Initiate the transport of the mobile phase, pushing the sample

through the chromatographic column. Compounds in the sample will separate based on differences in affinity and interactions with the stationary phase.

(6) Monitor and record data During the separation process, the detector monitors the signals of compounds and records detector output data.

(7) Analyze results Analyze the separated compounds based on the detector's output data. Typically, chromatograms are generated for identification and quantification using retention times and peak shapes.

(8) Cleaning and maintenance After analysis, perform system cleaning and maintenance, including column flushing, injector, detector cleaning, etc.

10.5.4.5 Applications

High-performance liquid chromatography (HPLC) is an essential chemical separation and analysis technique widely used in scientific research and industrial practices. It efficiently and accurately separates and detects compounds in mixtures, achieving precise separation of different compounds by adjusting various parameters. HPLC plays a crucial role in pharmaceuticals, food, environment, and other fields, with significant implications for drug component analysis, food safety testing, environmental pollution monitoring, etc.

10.5.5 Gas Chromatography

10.5.5.1 Principles of Method

Gas chromatography (**GC**) is based on the different distribution coefficients of substances in two relatively moving phases. As these substances move with the mobile phase, they undergo repeated distribution between the two phases, resulting in the efficient separation of components, subsequently entering the detector for determination.

10.5.5.2 Structure and Components of GC Instrument

Gas chromatographs mainly consist of a gas system, injection system, separation system, detection system, temperature control system, and recording system (Figure 10.8). The gas system includes a gas source, purification, drying tube, and carrier gas flow control device. Commonly used carrier gases include hydrogen, nitrogen, and argon, with a purity requirement of over 99.999%, and they should be chemically inert. The injection system consists of an injector and a vaporization chamber. As the core of the gas chromatograph, the separation system effectively separates components in the sample through a column box, chromatographic column, and temperature control components. The chromatographic column includes packed columns and capillary columns. The detection system converts the separated components in the chromatographic column into corresponding electrical signals according to changes in concentration. After amplification, the electrical signals are transmitted to the recorder, which draws the chromatogram.

10.5.5.3 Operating Procedure

The operating procedure is as follows:

(1) Instrument preparation Before using the gas chromatograph, ensure the

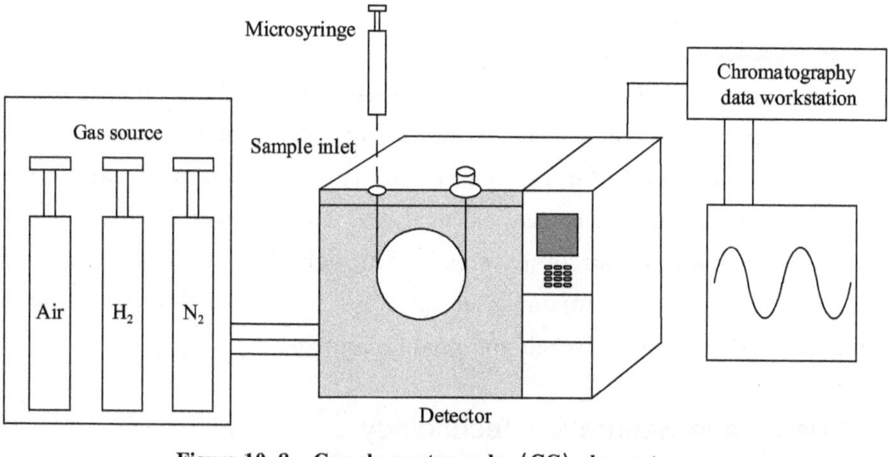

Figure 10.8 Gas chromatography (GC) chromatogram

instruments and materials are prepared for the experiment.

(2) Instrument activation Power on the gas chromatograph, ensuring the main unit and components' power indicators are illuminated. Perform system checks to verify the proper functioning of components, such as the injector, stabilizer, and detector. Set the gas chromatograph's temperature, flow rate, and sample volume parameters according to experimental requirements to ensure accurate experimental conditions.

(3) Sample handling Process the test sample appropriately according to experimental requirements. A sample needle extracts a specific sample volume and introduces it into the chromatographic column through the injector.

(4) Separation analysis Depending on the sample and analysis requirements, set the chromatographic column's temperature, adjust the carrier gas's flow rate and pressure, choose an appropriate detector, and configure corresponding parameters. Initiate the data acquisition system of the gas chromatograph to begin recording the signal responses of the sample, monitoring and recording the occurrence of chromatographic peaks in real time.

(5) Results interpretation Identify and confirm the analysis results based on the relative retention times of separated chromatographic peaks and specific detector responses.

10.5.5.4 Main Factors Influencing Separation

The primary factors influencing separation include the stationary phase, carrier gas flow rate, chromatographic column length and diameter, and sample characteristics. The stationary phase's polarity, particle size, and surface activity directly impact the chromatographic column's molecular interactions and separation efficiency. The flow rate of the carrier gas directly affects the migration speed of molecules in the chromatographic column. An excessive flow rate may lead to incomplete separation, while a flow rate that is too slow can affect analysis time and sensitivity. The length and diameter of the chromatographic column influence the time and degree of molecular spreading in the column. Longer columns and smaller diameters typically provide better separation effects but may also increase analysis time.

10.5.5.5 Application Areas

Gas chromatography (GC) is a widely used analytical technique in chemistry, biochemistry, and environmental science. It utilizes gas as the mobile phase to separate and detect compounds in samples, offering high sensitivity, high resolution, and rapid analysis advantages. GC is crucial in pharmaceutical analysis, environmental pollution monitoring, food safety, petrochemicals, and other areas. Through GC technology, qualitative and quantitative analysis of trace compounds in samples is achievable, providing a reliable analytical method for scientific research and quality control.

10.6 Membrane Separation Technology

Membrane separation technology is an efficient, economical, and convenient separation technique. Employing natural or synthetically created thin cohesive layers, known as membranes, this technique relies on external energy or chemical gradients (concentration, temperature, pressure, and electric potential) to drive the separation, enrichment, and purification of solutes and solvents involving two or more components. Membranes can exhibit full or partial permeability and exist in solid, liquid, or gas states, ensuring their presence between identical or dissimilar fluid phases. Serving as a distinct phase with two interfaces in contact with the separated substances but non-miscible with them, membranes are categorized based on the driving force employed in the method. This classification includes techniques such as osmosis, liquid membrane, and pervaporation based on concentration differences; reverse osmosis, ultrafiltration, and microfiltration based on pressure differences; and electrodialysis based on electric potential differences. In the subsequent sections, we focus on the widely utilized liquid membrane extraction separation method in analytical chemistry.

10.6.1 Composition and Properties of Liquid Membranes

A liquid membrane, characterized by its thinness and immiscibility with the separated liquid, can be classified into water-in-oil (W/O, oil type) and oil-in-water (O/W, water type). Unlike solid membranes, liquid membranes offer advantages such as rapid mass transfer, excellent selectivity, high separation efficiency, and operational simplicity. Furthermore, liquid membranes can be either carrier-free or carrier-mediated. The latter involves the introduction of a flowing carrier, such as trioctylamine, β-diketones, or fatty acids, influencing the separation process based on the carrier's properties. Carrier-free liquid membranes rely on selective permeation, chemical reactions, extraction, and adsorption for separation.

A liquid membrane comprises a solvent, surfactant, flowing carrier, and membrane enhancer. Surfactants play a pivotal role in stabilizing the liquid membrane. Emulsion membranes are prepared by introducing a carefully proportioned mixture of membrane solvent, surfactant, flowing carrier, and membrane enhancer into a stirred vessel with constant temperature control and speed adjustment. The selection of surfactants depends on whether an oil or water type is desired, with Span 80 and polyamine preferred for oil type and saponin for water type. The membrane solvent's viscosity is crucial for maintaining membrane stability and serves as the substrate for constructing membranes, requiring solvents to have a certain viscosity. The water type uses water as the membrane solvent, and glycerol can be added as the membrane enhancer. The oil type uses kerosene, neutral oil, and diesel as the membrane solvents. In carrier-mediated liquid membrane separation, the flowing carrier is essential for achieving separation and can be categorized into ionic and non-ionic types. Ionic carriers are divided into positive and negative charges, and it is generally believed that nonionic carriers are better than ionic carriers. Different carriers are selected based on the different separation objects. Generally, those suitable for extractants can be used as flow carriers, such as carboxylic acid and trioctyl, amines, oximes, and macrocyclic polyethers. Various additives are introduced to enhance liquid membrane stability.

A critical step in preparing supported liquid membranes involves immersing the liquid membrane solution containing the carrier into the pores of the support material, often inert polymer films such as polysulfone, polytetrafluoroethylene, or polypropylene films.

10.6.2 Separation Principles and Processes

Liquid membrane extraction involves three liquid phases: the sample solution, the extracting liquid, and the thin membrane between them. The separation principles resemble liquid-liquid extraction and counter-extraction. The target components, transitioning from the sample solution to the liquid membrane, represent extraction, while their subsequent movement from the liquid membrane to the receiving liquid phase represents counter-extraction. In essence, liquid membrane extraction amalgamates the advantages of liquid-liquid extraction by eliminating fundamental interferences found in dialysis processes, offering efficiency, speed, simplicity, and automation potential. The basic principle involves a porous thin film infused with an organic solvent immiscible with water, separating the aqueous solution into the extract and to-be-extracted phases. The phase connected to the flowing sample aqueous solution is the extracted phase, and the stationary phase is the extracted phase. The ions from the sample solution enter the to-be-extracted phase, forming neutral molecules in an activated state. These neutral molecules diffuse into and adsorb onto the liquid membrane, further diffusing into the extraction phase. Once in the extraction phase, these neutral molecules, influenced by the chemical conditions of the extraction phase, decompose into ions in a deactivated state, irreversible, and unable to return to the liquid membrane. The result is that the substance ions in the extracted phase enter the extraction phase through a liquid membrane. Figure 10.9 illustrates how the anions, amino groups, and metal ions in a water solution undergo extraction from the source phase into the extracting phase through a liquid membrane. As these ions flow into the extracting phase, they interact with the corresponding reagents introduced into it, namely H^+, OH^-, and ligands. This interaction leads to the formation of neutral molecules or coordination compounds entering the liquid membrane layer. Subsequently, these entities further diffuse into the extracting phase. Upon entering the extracting phase, they change chemical conditions influenced by adding acids, bases, or complexing agents. This dissociates the formed neutral molecules or coordination compounds into their original forms: acidic ions, amino ions, and metal ions, which are irreversibly retained in the extracting phase.

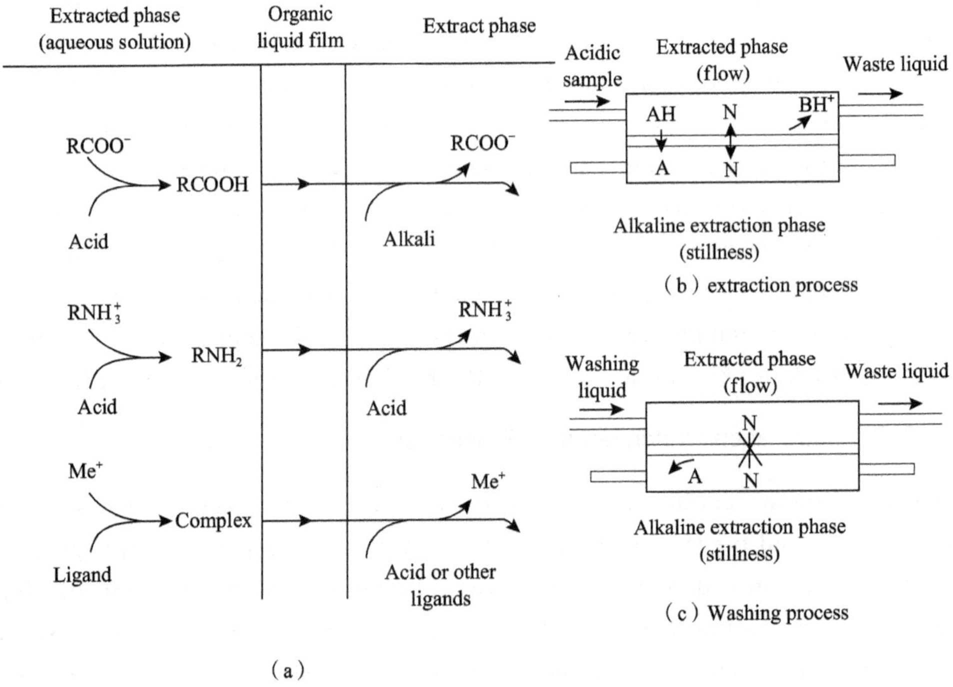

(a) Liquid membrane extraction of anions, amines, and metal ions;
(b)-(c) Separation of anions (A), cations (BH^+), and neutral ions (N) in the liquid membrane.

Figure 10.9 Diagram illustrating liquid membrane extraction of ions

10.6.3 Factors Influencing Liquid Membrane Extraction

In liquid membrane extraction separation, the separated substance in the flowing phase's aqueous solution can only enter the liquid membrane by transforming into an activated state (i. e., neutral molecules). Therefore, enhancing the selectivity of liquid membrane extraction separation technology depends on improving the ability of the separated substance to transition from a non-activated state to an activated state. This must be achieved without causing interference substances or other undesired materials to undergo activation. Consequently, the influencing factors include the following.

10.6.3.1 Acidity and medium

Adjusting the solution's pH extracts various substances based on their pK values. For example, extracting anionic species involves setting the pH to an acidic level, where anions and hydrogen ions combine to form corresponding acid molecules, facilitating their passage through the liquid membrane into the extraction phase. The cations exit the system with the aqueous solution. Acid molecules entering the extraction phase encounter an alkaline environment, releasing anions through interaction with surrounding hydroxide ions. Neutral molecules freely move between the two sides of the liquid membrane, entering the washing solution during the rinsing process. The result is that the anions in the aqueous solution selectively enter the extracting phase from the extracted phase. The situation is identical for cations, but the conditions are opposite. It is important to emphasize that the purpose of adjusting the pH is different for the extracted phase and the extracting phase. For the former, it is to convert the extracted substance from a non-activated state to an activated state, while for the latter, it is the opposite, transforming from an activated state to a non-activated state.

10.6.3.2 Polarity of organic liquid in the liquid membrane

Altering the polarity of the organic liquid in the liquid membrane enhances the extraction efficiency for substances with different polarities. The size of the liquid membrane's polarity directly influences the distribution coefficient of the substances within it. A closer match in polarity results in a larger distribution coefficient, making activated-state substances more prone to diffusion into the liquid membrane. Otherwise, even if the extracted substance forms neutral molecules in the aqueous phase and is activated, due to the significant polarity difference, it still cannot effectively enter the organic liquid membrane layer, affecting the extraction efficiency.

10.6.4 Applications

Using surfactants such as P507, dibutyl sulfoxide, liquid paraffin, and kerosene, liquid membranes can enrich rare earth cerium groups at concentrations as low as 10^{-9} g. Polysulfone P204 is used as a liquid membrane for separating Eu^{3+}. Liquid membrane extraction methods find widespread application in separating and enriching environmental samples, including isolating trace organic amines from atmospheric samples, separating copper ions and cobalt ions in water, and determining acidic pesticides in water bodies.

Besides the liquid membrane extraction method discussed above, analytical chemistry employs other separation techniques such as centrifugation, density gradient, dialysis, and electrodialysis.

Key Glossary

Chromatographic separation, commonly called chromatography, is also known as layer analysis and color layer method. Chromatography is a multi-stage separation technique based on the minute differences in the distribution coefficients of the substances being separated in two phases (stationary and mobile phases).

Extraction chromatography, a liquid-phase distribution chromatography combining solvent extraction with chromatographic separation techniques, is typically conducted in a column format known as column chromatography.

Thin-layer chromatography involves a liquid-phase chromatographic method with an adsorbent thin layer fixed on a planar support (such as a glass plate).

Paper chromatography uses filter paper as the carrier, with the stationary phase being the water absorbed on the filter paper.

High-performance liquid chromatography (HPLC) involves separating components dissolved in the mobile phase as they interact with the stationary phase. This interaction, which can be adsorption, distribution, exclusion, or affinity, varies in strength and size, resulting in different retention times within the stationary phase.

Gas chromatography (GC) is based on the different distribution coefficients of substances in two relatively moving phases. As these substances move with the mobile phase, they undergo repeated distribution between the two phases, resulting in the efficient separation of components, subsequently entering the detector for determination.

Membrane separation technology is an efficient, economical, and convenient separation technique. Employing natural or synthetically created thin cohesive layers, known as membranes, this technique relies on external energy or chemical gradients (concentration, temperature, pressure, and electric potential) to drive the separation, enrichment, and purification of solutes and solvents involving two or more components.

Ion exchange separation is a technique that utilizes the exchange of anions and cations to achieve separation.

Precipitation separation is a method of using a precipitation reaction, and the principle is to achieve separation according to the solubility of different substances in solvents.

Extraction separation is a liquid-liquid extraction method that utilizes the different solubility of substances in water.

Essential Exercises

Review Questions

1. What are the differences in stationary phases and distribution mechanisms among several chromatographic separation methods (paper chromatography, thin-layer chromatography, and extraction chromatography)?

2. How to perform quantitative measurements using paper chromatography and thin-layer chromatography?

3. What is the first requirement that must be met before a chemical or sample can be used in liquid chromatography (LC)?

4. What role does the mobile phase play in LC regarding analyte retention? How is this different from the role of the mobile phase in gas chromatography (GC)?

5. What general characteristics should a compound have if it is to be analyzed by gas chromatography?

6. Discuss how GC can be used for chemical separation and measurement.

7. Describe the separation principle of the membrane separation method.

8. Please describe what mixtures are suitable for ion exchange separation.

9. What is "precipitation separation" commonly? What are the similarities and differences?

10. Describe the principle and precautions of "extraction separation".

Cumulative Problems

1. Suppose a mixed solution containing components A and B, with $R_f = 0.40$ for component A and $R_f = 0.60$ for component B. Suppose the length of the filter paper used for chromatography is 20 cm. What is the maximum distance between the center of the spots after chromatographic separation of components A and B?

2. Injection of ethane, propane, butane, pentane, hexane, heptane, and octane on a squalene column under isothermal conditions gives retention times of 3.51 min, 4.98 min, 7.30 min, 11.00 min, 16.85 min, 26.12 min, and 40.8 min, respectively. Air, which is nonretained, elutes at 1.00 min. An injection of benzene, butanol, nitropropane, pyridine, and 2-pentanone under the same conditions gives retention times of 20.00 min, 6.50 min, 4.10 min, 10.05 min, and 10.35 min. Under these conditions, what are the Kovats retention indexes for benzene, butanol, nitropropane, pyridine, and 2-pentanone?

3. The baseline width measured in Problem 8 for phenytoin on the first column in solvent B was 0.46 min at a flow rate of 1.0 $mL \cdot min^{-1}$. The baseline width measured for phenytoin on the second column under the same conditions was 2.16 min. What is the number of theoretical plates and plate height for each of these columns? Is the first or second column more efficient in this separation?

Appendixes

Appendix 1 Relative Atomic Masses

Symbol	Element	Atomic weight	Symbol	Element	Atomic weight
Ac	Actinium	227.0	Er	Erbium	167.3
Ag	Silver	107.9	^{252}Es	Einsteinium	254.0
Al	Aluminium	26.98	Eu	Europium	152.0
^{243}Am	Americium	243.1	F	Fluorine	19.00
Ar	Argon	39.95	Fe	Iron	55.85
As	Arsenic	74.92	^{257}Fm	Fermium	257.1
^{210}At	Astatine	210.0	^{223}Fr	Francium	223.0
Au	Gold	197.0	Ga	Gallium	69.72
B	Boron	10.81	Gd	Gadolinium	157.3
Ba	Barium	137.3	Ge	Germanium	72.61±2
Be	Beryllium	9.012	H	Hydrogen	1.008
Bi	Bismuth	209.0	He	Helium	4.003
^{247}Bk	Berkelium	247.1	Hf	Hafnium	178.5
Br	Bromine	79.9	Hg	Mercury	200.6
C	Carbon	12.01	Ho	Holmium	164.9
Ca	Calcium	40.08	I	Iodine	126.9
Cd	Cadmium	112.4	In	Indium	114.8
Ce	Cerium	140.1	Ir	Iridium	192.2
^{252}Cf	Californium	252.1	K	Potassium	39.10
Cl	Chlorine	35.45	Kr	Krypton	83.80
^{247}Cm	Curium	247.1	La	Lamhanum	138.9
Co	Cobalt	58.93	Li	Lithium	6.941
Cr	Chromium	52.00	Lr	Lawrencium	257.0
Cs	Caesium	132.9	Lu	Lutetium	175.0
Cu	Copper	63.55	Md	Mendelevium	256.1
Dy	Dysprosium	162.5	Mg	Magnesium	24.31

Continued table

Symbol	Element	Atomic weight	Symbol	Element	Atomic weight
Mn	Manganese	54.94	Ru	Ruthenium	101.1
Mo	Molybdenum	95.94	S	Sulfur	32.07
N	Nitrogen	14.01	Sb	Antimony	121.8
Na	Sodium	22.99	Sc	Scandium	44.96
Nb	Niobium	92.91	Se	Selenium	78.96±3
Nd	Neodymium	144.2	Si	Silicon	28.09
Ne	Neon	20.18	Sm	Samarium	150.4
^{59}Ni	Nickel	58.69	Sn	Tin	118.7
No	Nobelium	254.0	Sr	Strontium	87.62
Np	Neptunium	237.0	Ta	Tantalum	180.9
O	Oxygen	16.00	Tb	Terbium	158.9
Os	Osmium	190.2	Tc	Technetium	98.91
P	Phosphorus	30.97	Te	Tellurium	127.6
^{231}Pa	Protactinium	231.0	Th	Thorium	232.0
Pb	Lead	207.2	Ti	Titanium	47.88±2
Pd	Palladium	106.4	Tl	Thallium	204.4
^{145}Pm	Promethium	144.9	Tm	Thulium	168.9
^{210}Po	Polonium	210.0	U	Uranium	238.0
Pr	Praseodymium	140.9	V	Vanadium	50.94
Pt	Platinum	195.1	W	Tungsten	183.9
^{239}Pu	Plutonium	244.0	Xe	Xenon	131.1
Ra	Radium	226.0	Y	Yttrium	88.91
Rb	Rubidium	85.47	Yb	Ytterbium	173.0
Re	Rhenium	186.2	Zn	Zinc	65.39±2
Rh	Rhodium	102.9	Zr	Zirconium	91.22
^{222}Rn	Radon	222.0			

Appendix 2 Formula Weights

Compound	Formula weights	Compound	Formula weights	Compound	Formula weights
Ag_3AsO_4	462.53	$CaSO_4$	136.14	NH_4Cl	53.49
AgBr	187.78	$CdCO_3$	172.41	H_3BO_3	61.83
AgCl	143.32	$CdCl_2$	183.33	CH_3COOH	60.05
AgCN	133.89	CdS	144.47	HCl	36.46
Ag_2CrO_4	331.73	$Ce(SO_4)_2$	332.24	H_2CO_3	62.03
AgI	234.77	$CoCl_2$	129.84	$H_2C_2O_4$	90.04
$AgNO_3$	169.87	CoS	90.99	HF	20.01
AgSCN	165.95	$CoSO_4$	154.99	HI	127.91
$AlCl_3$	133.33	$CrCl_3$	158.36	HIO_3	175.61
Al_2O_3	101.96	Cr_2O_3	151.99	HNO_3	63.02
$Al(OH)_3$	78.00	CuCl	99.00	H_2O	18.02
$Al_2(SO_4)_3$	342.17	$CuCl_2$	134.45	H_2O_2	34.02
As_2O_3	197.84	CuI	190.45	H_3PO_4	97.99
As_2O_5	229.84	CuO	79.55	H_2S	34.08
As_2S_3	246.05	Cu_2O	143.09	H_2SO_4	98.09
$BaCO_3$	197.31	CuS	95.62	$HgCl_2$	271.50
BaC_2O_4	225.32	$CuSO_4$	159.62	Hg_2Cl_2	472.09
$BaCl_2$	208.25	$CO(NH_2)_2$	60.05	HgI_2	454.40
$BaCl_2 \cdot 2H_2O$	244.37	CO_2	44.01	HgO	216.59
$BaCrO_4$	253.32	$FeCl_3$	162.21	HgS	232.65
$Ba(OH)_2$	171.35	FeO	71.85	$HgSO_4$	296.67
$BaSO_4$	233.37	Fe_2O_3	159.69	KBr	119.00
$CaCl_2$	110.99	$Fe(OH)_3$	106.88	$KBrO_3$	167.00
$CaCO_3$	100.09	$NH_4Fe(SO_4)_2 \cdot 12H_2O$	486.22	KCl	74.55
CaC_2O_4	128.10	$FeSO_4 \cdot (NH_4)_2SO_4 \cdot 6H_2O$	392.14	K_2CO_3	138.21
CaO	56.08	H_3AsO_3	125.94	KCN	65.12
$Ca(OH)_2$	74.10	H_3AsO_4	141.94	K_2CrO_4	194.19
$Ca_3(PO_4)_2$	310.18	NH_3	17.03	$KHC_2O_4 \cdot H_2C_2O_4 \cdot 2H_2O$	254.19

Continued table

Compound	Formula weights	Compound	Formula weights	Compound	Formula weights
KI	166.00	$MnSO_4 \cdot 4H_2O$	223.06	NH_4VO_3	116.98
$KHC_8H_4O_4$(KHP)	204.22	Na_2CO_3	105.99	$NiCl_2 \cdot 6H_2O$	237.69
KIO_3	214.00	$Na_2C_2O_4$	134.00	$NiSO_4 \cdot 7H_2O$	280.87
$KMnO_4$	158.03	CH_3COONa(NaAc)	82.03	$PbSO_4$	303.27
KNO_3	101.10	NaCl	58.44	P_2O_5	141.91
K_2O	94.20	$NaHCO_3$	84.01	$PbCl_2$	278.10
KOH	56.11	$Na_2HPO_4 \cdot 12H_2O$	358.14	$PbCO_3$	267.21
K_2SO_4	174.27	$Na_2H_2Y \cdot 2H_2O$	372.24	$Pb_3(PO_4)_2$	811.54
$K_2Cr_2O_7$	294.19	$NaNO_3$	85.00	OH^-	17.01
$MgCO_3$	84.32	NaOH	40.00	SO_2	64.07
$MgCl_2$	95.22	Na_2SO_4	142.05	SiF_4	104.08
MgC_2O_4	112.33	$Na_2S_2O_3$	158.12	SiO_2	60.08
$MgNH_4PO_4$	137.32	$Na_2B_4O_7 \cdot 10H_2O$	381.42	$SnCl_2$	189.60
MgO	40.31	$Na_2S_2O_3 \cdot 5H_2O$	248.20	$SnCl_4$	260.50
$Mg(OH)_2$	58.33	$(NH_4)_2CO_3$	96.09	$SrCO_3$	147.63
$Mg_2P_2O_7$	222.55	$(NH_4)_2C_2O_4$	124.10	$SrCrO_4$	203.62
$MgSO_4 \cdot 7H_2O$	246.49	NH_4HCO_3	79.06	$SrSO_4$	183.6
$MnCO_3$	114.95	$(NH_4)_2MoO_4$	196.01	$ZnCO_3$	125.39
$MnCl_2 \cdot 4H_2O$	197.91	NH_4NO_3	80.04	$ZnCl_2$	136.29
MnO_2	86.94	$(NH_4)_2HPO_4$	132.06	$ZnSO_4$	161.46
MnS	87.01	$(NH_4)_2SO_4$	132.15	$ZnSO_4 \cdot 7H_2O$	287.57

Appendix 3 Relative Densities and Concentrations of the Common Acids and Bases

Acids

Relative density (15 ℃)	The concentration of hydrochloric acid		The concentration of nitric acid		The concentration of sulphuric acid	
	$g \cdot 100\ g^{-1}$	$mol \cdot L^{-1}$	$g \cdot 100\ g^{-1}$	$mol \cdot L^{-1}$	$g \cdot 100\ g^{-1}$	$mol \cdot L^{-1}$
1.02	4.13	1.15	3.70	0.6	3.1	0.3
1.04	8.16	2.3	7.26	1.2	6.1	0.6

Continued table

Relative density (15 ℃)	The concentration of hydrochloric acid		The concentration of nitric acid		The concentration of sulphuric acid	
	g • 100 g^{-1}	mol • L^{-1}	g • 100 g^{-1}	mol • L^{-1}	g • 100 g^{-1}	mol • L^{-1}
1.05	10.2	2.9	9.0	1.5	7.4	0.8
1.06	12.2	3.5	10.7	1.8	8.8	0.9
1.08	16.2	4.8	13.9	2.4	11.6	1.3
1.10	20.0	6.0	17.1	3.0	14.4	1.6
1.12	23.8	7.3	20.2	3.6	17.0	2.0
1.14	27.7	8.7	23.3	4.2	19.9	2.3
1.15	29.6	9.3	24.8	4.5	20.9	2.5
1.19	37.2	12.2	30.9	5.8	26.0	3.2
1.20			32.3	6.2	27.3	3.4
1.25			39.3	7.9	33.4	4.3
1.30			47.5	9.8	39.2	5.2
1.35			55.8	12.0	44.8	6.2
1.40			65.3	14.5	50.1	7.2
1.42			69.8	15.7	52.2	7.6
1.45					55.0	8.2
1.50					59.8	9.2
1.55					64.3	10.2
1.60					68.7	11.2
1.65					73.0	12.3
1.70					77.2	13.4
1.84					95.6	18.0

Bases

Relative density (15 ℃)	The concentration of aqueous ammonia		The concentration of sodium hydroxide		The concentration of potassium hydroxide	
	g • 100 g^{-1}	mol • L^{-1}	g • 100 g^{-1}	mol • L^{-1}	g • 100 g^{-1}	mol • L^{-1}
0.88	35.0	18.0				
0.90	28.3	15.0				
0.91	25.0	13.4				
0.92	21.8	11.8				
0.94	15.6	8.6				
0.96	9.9	5.6				
0.98	4.8	2.8				

Continued table

Relative density (15 ℃)	The concentration of aqueous ammonia		The concentration of sodium hydroxide		The concentration of potassium hydroxide	
	$g \cdot 100\ g^{-1}$	$mol \cdot L^{-1}$	$g \cdot 100\ g^{-1}$	$mol \cdot L^{-1}$	$g \cdot 100\ g^{-1}$	$mol \cdot L^{-1}$
1.05			4.5	1.25	5.5	1.0
1.10			9.0	2.5	10.9	2.1
1.15			13.5	3.9	16.1	3.3
1.20			18.0	5.4	21.2	4.5
1.25			22.5	7.0	26.1	5.8
1.30			27.0	8.8	30.9	7.2
1.35			31.8	10.7	35.5	8.5

Appendix 4 Dissociation Constants of the Common Acids and Bases (25 ℃)

Acid/Base	Formula	$K_a^\ominus(K_b^\ominus)$	$pK_a^\ominus(pK_b^\ominus)$
Arsenic acid	H_3AsO_4	$6.3\times10^{-3}(K_{a_1}^\ominus)$	2.20
		$1.0\times10^{-7}(K_{a_2}^\ominus)$	7.00
		$3.2\times10^{-12}(K_{a_3}^\ominus)$	11.50
Arsenious acid	H_3AsO_3	$6.0\times10^{-10}(K_{a_1}^\ominus)$	9.22
		$3.0\times10^{-14}(K_{a_2}^\ominus)$	13.52
Boronic acid	H_3BO_3	5.8×10^{-10}	9.24
Pyroboric acid	$H_2B_4O_7$	$1.0\times10^{-4}(K_{a_1}^\ominus)$	4.00
		$1.0\times10^{-9}(K_{a_2}^\ominus)$	9.00
Carbonic acid	H_2CO_3	$4.2\times10^{-7}(K_{a_1}^\ominus)$	6.38
		$5.6\times10^{-11}(K_{a_2}^\ominus)$	10.25
Hydrocyanic acid	HCN	6.2×10^{-10}	9.21
Chromic acid	H_2CrO_4	$1.8\times10^{-1}(K_{a_1}^\ominus)$	0.74
		$3.2\times10^{-7}(K_{a_2}^\ominus)$	6.50
Hydrofluoric acid	HF	6.6×10^{-4}	3.18
Nitrous acid	HNO_2	5.1×10^{-4}	3.29

Continued table

Acid/Base	Formula	$K_a^{\ominus}(K_b^{\ominus})$	$pK_a^{\ominus}(pK_b^{\ominus})$
Hydrogen peroxide	H_2O_2	2.2×10^{-12}	11.66
Phosphoric acid	H_3PO_4	$7.6\times10^{-3}(K_{a_1}^{\ominus})$	2.12
		$6.3\times10^{-8}(K_{a_2}^{\ominus})$	7.20
		$4.4\times10^{-13}(K_{a_3}^{\ominus})$	12.36
Pyrophosphoric acid	$H_4P_2O_7$	$3.0\times10^{-2}(K_{a_1}^{\ominus})$	1.52
		$4.4\times10^{-3}(K_{a_2}^{\ominus})$	2.36
		$2.5\times10^{-7}(K_{a_3}^{\ominus})$	6.60
		$5.6\times10^{-10}(K_{a_4}^{\ominus})$	9.25
Phosphorous acid	H_3PO_3	$5.0\times10^{-2}(K_{a_1}^{\ominus})$	1.30
		$2.5\times10^{-7}(K_{a_2}^{\ominus})$	6.60
Hydrosulphuric acid	H_2S	$1.3\times10^{-7}(K_{a_1}^{\ominus})$	6.88
		$7.1\times10^{-15}(K_{a_2}^{\ominus})$	14.15
Sulfuric acid	H_2SO_4	$1.0\times10^{-2}(K_{a_2}^{\ominus})$	1.99
Sulphurous acid	H_2SO_3	$1.3\times10^{-2}(K_{a_1}^{\ominus})$	1.90
		$6.3\times10^{-8}(K_{a_2}^{\ominus})$	7.20
Metasilicic acid	H_2SiO_3	$1.7\times10^{-10}(K_{a_1}^{\ominus})$	9.77
		$1.6\times10^{-12}(K_{a_2}^{\ominus})$	11.80
Formic acid	HCOOH	1.8×10^{-4}	3.74
Acetic acid	CH_3COOH	1.8×10^{-5}	4.74
Lactic acid	$CH_3CHOHCOOH$	1.4×10^{-4}	3.86
Benzoic acid	C_6H_5COOH	6.2×10^{-5}	4.21
Oxalic acid	$H_2C_2O_4$	$5.9\times10^{-2}(K_{a_1}^{\ominus})$	1.22
		$6.4\times10^{-5}(K_{a_2}^{\ominus})$	4.19
Phthalic acid	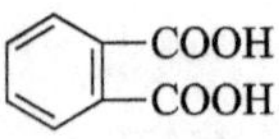	$1.1\times10^{-3}(K_{a_1}^{\ominus})$	2.95
		$3.9\times10^{-6}(K_{a_2}^{\ominus})$	5.41

Continued table

Acid/Base	Formula	$K_a^\ominus(K_b^\ominus)$	$pK_a^\ominus(pK_b^\ominus)$
Citric acid	$H_3OHC_6H_4O_6$	$7.1\times10^{-4}(K_{a_1}^\ominus)$	3.15
		$1.7\times10^{-5}(K_{a_2}^\ominus)$	4.76
		$9.1\times10^{-7}(K_{a_3}^\ominus)$	6.04
Phenol	C_6H_5OH	1.1×10^{-10}	9.95
Ethylenediaminetetra-acetic acid (EDTA)	H_6Y^{2+}	$1.3\times10^{-1}(K_{a_1}^\ominus)$	0.89
	H_5Y^+	$2.5\times10^{-2}(K_{a_2}^\ominus)$	1.60
	H_4Y	$1.0\times10^{-2}(K_{a_3}^\ominus)$	2.00
	H_3Y^-	$2.1\times10^{-3}(K_{a_4}^\ominus)$	2.67
	H_2Y^{2-}	$6.9\times10^{-7}(K_{a_5}^\ominus)$	6.16
	HY^{3-}	$5.5\times10^{-11}(K_{a_6}^\ominus)$	10.26
Ammonium hydroxide	NH_3	1.8×10^{-5}	4.74
Hydroxylamine	NH_2OH	9.1×10^{-6}	8.04
Methylamine	CH_3NH_2	4.2×10^{-4}	3.38
Ethylamine	$C_2H_5NH_2$	5.6×10^{-4}	3.25
Dimethylamine	$(CH_3)_2NH$	1.2×10^{-4}	3.93
Diethylamine	$(C_2H_5)_2NH$	1.3×10^{-3}	2.89
Ethanolamine	$HOCH_2CH_2NH_2$	3.2×10^{-5}	4.50
Triethanolamine	$(HOCH_2CH_2)_3N$	5.8×10^{-7}	6.24
Hexamethylene tetramine	$(CH_2)_6N_4$	1.4×10^{-9}	8.85
Ethanediamine	$H_2NHC_2CH_2NH_2$	$8.5\times10^{-5}(K_{b_1}^\ominus)$	4.07
		$7.1\times10^{-8}(K_{b_2}^\ominus)$	7.15
Pyridine	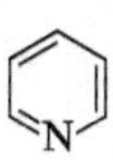	1.7×10^{-9}	8.77
Byophenanthroline	$C_{12}H_8N_2$	1.1×10^{-5}	4.96

Appendix 5 Solubility Product Constants of Insoluble Compounds (18-25 ℃)

Compound (molecular formula)	$K_{sp}^{\ominus}$	$pK_{sp}^{\ominus}$	Compound (molecular formula)	$K_{sp}^{\ominus}$	$pK_{sp}^{\ominus}$
Ag_3AsO_4	1.0×10^{-22}	22.0	$Al(OH)_3$	4.57×10^{-33}	32.3
$AgBr$	5.0×10^{-13}	12.3	$AlPO_4$	6.3×10^{-19}	18.2
$AgBrO_3$	5.50×10^{-5}	4.26	Al_2S_3	2.0×10^{-7}	6.7
$AgCl$	1.8×10^{-10}	9.74	$Au(OH)_3$	5.5×10^{-46}	45.26
$AgCN$	1.2×10^{-16}	15.9	$AuCl_3$	3.2×10^{-25}	24.5
Ag_2CO_3	8.1×10^{-12}	11.1	AuI_3	1.0×10^{-46}	46.0
$Ag_2C_2O_4$	3.5×10^{-11}	10.5	$Ba_3(AsO_4)_2$	8.0×10^{-51}	50.1
Ag_2CrO_4	1.2×10^{-12}	11.9	$BaCO_3$	5.1×10^{-9}	8.29
$Ag_2Cr_2O_7$	2.0×10^{-7}	6.7	BaC_2O_4	1.6×10^{-7}	6.79
AgI	8.3×10^{-17}	16.1	$BaCrO_4$	1.2×10^{-10}	9.93
$AgIO_3$	3.1×10^{-8}	7.51	$Ba_3(PO_4)_2$	3.4×10^{-23}	22.4
$AgOH$	2.0×10^{-8}	7.71	$BaSO_4$	1.1×10^{-10}	9.96
Ag_2MoO_4	2.8×10^{-12}	11.6	BaS_2O_3	1.6×10^{-5}	4.79
Ag_3PO_4	1.4×10^{-16}	15.8	$BaSeO_3$	2.7×10^{-7}	6.57
Ag_2S	6.3×10^{-50}	49.2	$BaSeO_4$	3.5×10^{-8}	7.46
$AgSCN$	1.0×10^{-12}	12.0	$Be(OH)_2$	1.6×10^{-22}	21.8
Ag_2SO_3	1.5×10^{-14}	13.8	$BiAsO_4$	4.4×10^{-10}	9.36
Ag_2SO_4	1.4×10^{-5}	4.84	$Bi_2(C_2O_4)_3$	3.98×10^{-36}	35.4
Ag_2Se	2.0×10^{-64}	63.7	$Bi(OH)_3$	4.0×10^{-31}	30.4
Ag_2SeO_3	1.0×10^{-15}	15.0	$BiPO_4$	1.26×10^{-23}	22.9
Ag_2SeO_4	5.7×10^{-8}	7.25	$CaCO_3$	2.8×10^{-9}	8.54
$AgVO_3$	5.0×10^{-7}	6.3	$CaC_2O_4 \cdot H_2O$	4.0×10^{-9}	8.39
Ag_2WO_4	5.5×10^{-12}	11.3	CaF_2	2.7×10^{-11}	10.6

Continued table

Compound (molecular formula)	$K_{sp}^{\ominus}$	$pK_{sp}^{\ominus}$	Compound (molecular formula)	$K_{sp}^{\ominus}$	$pK_{sp}^{\ominus}$
$CaMoO_4$	4.17×10^{-8}	7.38	$CrPO_4\cdot4H_2O$(green)	2.4×10^{-23}	22.6
$Ca(OH)_2$	5.5×10^{-6}	5.26	$CrPO_4\cdot4H_2O$(purple)	1.0×10^{-17}	17.0
$Ca_3(PO_4)_2$	2.0×10^{-29}	28.7	CuBr	5.3×10^{-9}	8.28
$CaSO_4$	3.16×10^{-7}	5.04	CuCl	1.2×10^{-6}	5.92
$CaSiO_3$	2.5×10^{-8}	7.6	CuCN	3.2×10^{-20}	19.5
$CaWO_4$	8.7×10^{-9}	8.06	$CuCO_3$	2.34×10^{-10}	9.63
$CdCO_3$	5.2×10^{-12}	11.3	CuI	1.1×10^{-12}	12.0
$CdC_2O_4\cdot3H_2O$	9.1×10^{-8}	7.04	$Cu(OH)_2$	4.8×10^{-20}	19.3
$Cd_3(PO_4)_2$	2.5×10^{-33}	32.6	$Cu_3(PO_4)_2$	1.3×10^{-37}	36.9
CdS	8.0×10^{-27}	26.1	Cu_2S	2.5×10^{-48}	47.6
CdSe	6.31×10^{-36}	35.2	Cu_2Se	1.58×10^{-61}	60.8
$CdSeO_3$	1.3×10^{-9}	8.89	CuS	6.3×10^{-36}	35.2
CeF_3	8.0×10^{-16}	15.1	CuSe	7.94×10^{-49}	48.1
$CePO_4$	1.0×10^{-23}	23.0	$Dy(OH)_3$	1.4×10^{-22}	21.9
$Co_3(AsO_4)_2$	7.6×10^{-29}	28.1	$Er(OH)_3$	4.1×10^{-24}	23.4
$CoCO_3$	1.4×10^{-13}	12.8	$Eu(OH)_3$	8.9×10^{-24}	23.05
CoC_2O_4	6.3×10^{-8}	7.2	$FeAsO_4$	5.7×10^{-21}	20.2
$Co(OH)_2$(blue)	6.31×10^{-15}	14.2	$FeCO_3$	3.2×10^{-11}	10.5
$Co(OH)_2$(pink, precipitate)	1.58×10^{-15}	14.8	$Fe(OH)_2$	8.0×10^{-16}	15.1
$Co(OH)_2$(pink, aging)	2.0×10^{-16}	15.7	$Fe(OH)_3$	4.0×10^{-38}	37.4
$CoHPO_4$	2.0×10^{-7}	6.69	$FePO_4$	1.3×10^{-22}	21.9
$Co_3(PO_4)_3$	2.0×10^{-35}	34.7	FeS	6.3×10^{-18}	17.2
$CrAsO_4$	7.7×10^{-21}	20.1	$Ga(OH)_3$	7.0×10^{-36}	35.2
$Cr(OH)_3$	6.3×10^{-31}	30.2	$GaPO_4$	1.0×10^{-21}	21.0

Continued table

Compound (molecular formula)	$K_{sp}^{\ominus}$	$pK_{sp}^{\ominus}$	Compound (molecular formula)	$K_{sp}^{\ominus}$	$pK_{sp}^{\ominus}$
$Gd(OH)_3$	1.8×10^{-23}	22.7	$MgCO_3$	3.5×10^{-8}	7.46
$Hf(OH)_4$	4.0×10^{-26}	25.4	$MgCO_3\cdot3H_2O$	2.14×10^{-5}	4.67
Hg_2Br_2	5.6×10^{-23}	22.2	$Mg(OH)_2$	1.8×10^{-11}	10.7
Hg_2Cl_2	1.3×10^{-18}	17.9	$Mg_3(PO_4)_2\cdot8H_2O$	6.31×10^{-26}	25.2
HgC_2O_4	1.0×10^{-7}	7.0	$Mn_3(AsO_4)_2$	1.9×10^{-29}	28.7
Hg_2CO_3	8.9×10^{-17}	16.1	$MnCO_3$	1.8×10^{-11}	10.7
$Hg_2(CN)_2$	5.0×10^{-40}	39.3	$Mn(IO_3)_2$	4.37×10^{-7}	6.36
Hg_2CrO_4	2.0×10^{-9}	8.7	$Mn(OH)_4$	1.9×10^{-13}	12.7
Hg_2I_2	4.5×10^{-29}	28.4	MnS(pink)	2.5×10^{-10}	9.6
HgI_2	2.82×10^{-29}	28.6	MnS(green)	2.5×10^{-13}	12.6
$Hg_2(IO_3)_2$	2.0×10^{-14}	13.7	$Ni_3(AsO_4)_2$	3.1×10^{-26}	25.5
$Hg_2(OH)_2$	2.0×10^{-24}	23.7	$NiCO_3$	6.6×10^{-9}	8.18
HgSe	1.0×10^{-59}	59	NiC_2O_4	4.0×10^{-10}	9.39
HgS(red)	4.0×10^{-53}	52.4	$Ni(OH)_2$	2.0×10^{-15}	14.7
HgS(black)	1.6×10^{-52}	51.8	$Ni_3(PO_4)_2$	5.0×10^{-31}	30.3
Hg_2WO_4	1.1×10^{-17}	17.0	α-NiS	3.2×10^{-19}	18.5
$Ho(OH)_3$	5.0×10^{-23}	22.3	β-NiS	1.0×10^{-24}	24.0
$In(OH)_3$	1.3×10^{-37}	36.9	γ-NiS	2.0×10^{-26}	25.7
$InPO_4$	2.3×10^{-22}	21.6	$Pb_3(AsO_4)_2$	4.0×10^{-36}	35.4
In_2S_3	5.7×10^{-74}	73.2	$PbBr_2$	4.0×10^{-5}	4.41
$La_2(CO_3)_3$	3.98×10^{-34}	33.4	$PbCl_2$	1.6×10^{-5}	4.79
$LaPO_4$	3.98×10^{-23}	22.4	$PbCO_3$	7.4×10^{-14}	13.1
$Lu(OH)_3$	1.9×10^{-24}	23.7	$PbCrO_4$	2.8×10^{-13}	12.6
$Mg_3(AsO_4)_2$	2.1×10^{-20}	19.7	PbF_2	2.7×10^{-8}	7.57

Continued table

Compound (molecular formula)	$K_{sp}^{\ominus}$	$pK_{sp}^{\ominus}$	Compound (molecular formula)	$K_{sp}^{\ominus}$	$pK_{sp}^{\ominus}$
$PbMoO_4$	1.0×10^{-13}	13.0	$Sn(OH)_4$	1.0×10^{-56}	56.0
$Pb(OH)_2$	1.2×10^{-15}	14.9	SnO_2	3.98×10^{-65}	64.4
$Pb(OH)_4$	3.2×10^{-66}	65.5	SnS	1.0×10^{-25}	25.0
$Pb_3(PO_4)_3$	8.0×10^{-43}	42.1	SnSe	3.98×10^{-39}	38.4
PbS	1.0×10^{-28}	28.0	$Sr_3(AsO_4)_2$	8.1×10^{-19}	18.1
$PbSO_4$	1.6×10^{-8}	7.79	$SrCO_3$	1.1×10^{-10}	9.96
PbSe	7.94×10^{-43}	42.1	$SrC_2O_4 \cdot H_2O$	1.6×10^{-7}	6.8
$PbSeO_4$	1.4×10^{-7}	6.84	SrF_2	2.5×10^{-9}	8.61
$Pd(OH)_2$	1.0×10^{-31}	31.0	$Sr_3(PO_4)_2$	4.0×10^{-28}	27.4
$Pd(OH)_4$	6.3×10^{-71}	70.2	$SrSO_4$	3.2×10^{-7}	6.49
PdS	2.03×10^{-58}	57.7	$SrWO_4$	1.7×10^{-10}	9.77
$Pm(OH)_3$	1.0×10^{-21}	21.0	$Tb(OH)_3$	2.0×10^{-22}	21.7
$Pr(OH)_3$	6.8×10^{-22}	21.2	$Te(OH)_4$	3.0×10^{-54}	53.5
$Pt(OH)_2$	1.0×10^{-35}	35.0	$Th(C_2O_4)_2$	1.0×10^{-22}	22.0
$Pu(OH)_3$	2.0×10^{-20}	19.7	$Th(IO_3)_4$	2.5×10^{-15}	14.6
$Pu(OH)_4$	1.0×10^{-55}	55.0	$Th(OH)_4$	4.0×10^{-45}	44.4
$RaSO_4$	4.2×10^{-11}	10.4	$Ti(OH)_3$	1.0×10^{-40}	40.0
$Rh(OH)_3$	1.0×10^{-23}	23.0	TlBr	3.4×10^{-6}	5.47
$Ru(OH)_3$	1.0×10^{-36}	36.0	TlCl	1.7×10^{-4}	3.76
Sb_2S_3	1.5×10^{-93}	92.8	Tl_2CrO_4	9.77×10^{-13}	12.0
ScF_3	4.2×10^{-18}	17.4	TlI	6.5×10^{-8}	7.19
$Sc(OH)_3$	8.0×10^{-31}	30.1	TlN_3	2.2×10^{-4}	3.66
$Sm(OH)_3$	8.2×10^{-23}	22.1	Tl_2S	5.0×10^{-21}	20.3
$Sn(OH)_2$	1.4×10^{-28}	27.9	$TlSeO_3$	2.0×10^{-39}	38.7

Continued table

Compound (molecular formula)	$K_{sp}^{\ominus}$	$pK_{sp}^{\ominus}$	Compound (molecular formula)	$K_{sp}^{\ominus}$	$pK_{sp}^{\ominus}$
$UO_2(OH)_2$	1.1×10^{-22}	22.0	$Zn(OH)_2$	2.09×10^{-16}	15.7
$VO(OH)_2$	5.9×10^{-23}	22.1	$Zn_3(PO_4)_2$	9.0×10^{-33}	32.0
$Y(OH)_3$	8.0×10^{-23}	22.1	α-ZnS	1.6×10^{-24}	23.8
$Yb(OH)_3$	3.0×10^{-24}	23.5	β-ZnS	2.5×10^{-22}	21.6
$Zn_3(AsO_4)_2$	1.3×10^{-28}	27.9	$ZrO(OH)_2$	6.3×10^{-49}	48.2
$ZnCO_3$	1.4×10^{-11}	10.8			

Appendix 6 Formation Constants of Some Complexes(18-25 ℃)

Complex	lg K_f	Complex	lg K_f
$[AgCl]^{3-}$	5.31	$[Co(SCN)_4]^{2-}$	3.0
$[AgBr]^{3-}$	9.2	$[Co(NH_3)_6]^{2+}$	5.11
$[Ag(CN)_2]^-$	21.1	$[Co(NH_3)_6]^{3+}$	35.2
$[Ag(SCN)]^-$	7.57	$[Co(C_2O_4)_3]^{4-}$	9.7
$[AgC_2O_4]^-$	2.41	$[Co(C_2O_4)_3]^{3-}$	−20
$[Ag(NH_3)_2]^+$	7.03	CrF_3	10.3
$[Ag(S_2O_3)_2]^{3-}$	13.5	$[Cr(SCN)_2]^+$	3.0
$[AlF_6]^{3-}$	19.8	$[CuCl_3]^{2-}$	5.7
$[Al(C_2O_4)_3]^{3-}$	16.3	$[CuBr_2]^-$	5.89
$[CdCl_4]^{2-}$	2.8	$[CuI_2]^-$	8.76
$[CdBr_4]^{2-}$	3.7	$[Cu(CN)_4]^{3-}$	30.3
$[CdI_4]^{2-}$	6.49	$[Cu(SCN)_2]^+$	5.18
$[Cd(CN)_4]^{2-}$	18.9	$[Cu(NH_2)_2]^+$	10.8
$[Cd(SCN)_4]^{2-}$	3.6	$[Cu(NH_3)_4]^{2+}$	12.67
$[Cd(NH_3)_4]^{2+}$	7.12	$[Cu(S_2O_3)_3]^{5-}$	13.8
$[Cd(S_2O_3)_2]^{2-}$	6.44	FeF_3	12.1
$[Co(CN)_4]^{2-}$	19.1	$[CuCl_3]^{2-}$	5.7

Continued table

Complex	lg K_f	Complex	lg K_f
$[CuBr_2]^-$	5.89	$[HgI_4]^{2-}$	29.8
$[CuI_2]^-$	8.76	$[Hg(CN)_4]^{2-}$	41.4
$[Cu(CN)_4]^{3-}$	30.3	$[Hg(SCN)_4]^{2-}$	21.2
$[Cu(SCN)_2]^-$	5.18	$[Hg(NH_3)_4]^{2+}$	19.4
$[Cu(NH_2)_2]^+$	10.8	$[Hg(S_2O_3)_4]^{6-}$	33.2
$[Cu(NH_3)_4]^{2+}$	12.67	$[Hg(C_2O_4)]^{2-}$	6.98
$[Cu(S_2O_3)_3]^{5-}$	13.8	$[Mg(C_2O_4)_2]^{2-}$	4.38
FeF_3	12.1	$[MnF]^+$	5.48
$[FeF_5]^{2-}$	15.8	$[Mn(C_2O_4)_3]^{4-}$	19.4
$[Fe(CN)_6]^{4-}$	35	$[PbBr_4]^{2-}$	3.0
$[Fe(CN)_6]^{3-}$	42	$[PbI_4]^{2-}$	4.47
$Fe(SCN)_3$	5.64	$[Pb(CN)_4]^{2-}$	10.0
$[Fe(SCN)_5]^{2-}$	6.4	$[Pb(S_2O_3)_3]^{4-}$	6.35
$[FeHPO_4]^+$	9.35	$[Pb(C_2O_4)_2]^{2-}$	6.54
$[Fe(C_2O_4)_3]^{4-}$	5.2	$[Zn(CN)_4]^{2-}$	16.7
$[Fe(C_2O_4)_3]^{3-}$	20.2	$[Zn(NH_3)_4]^{2+}$	9.46
$[HgCl_4]^{2-}$	16.2	$[Zn(C_2O_4)_3]^4$	8.15
$[HgBr_4]^{2-}$	21.0		

Appendix 7 Formation Constants of Metal Ions with Various Chelating Agents (18-25 ℃)

Metal ion	lg K					
	EDTA	CyDTA	DTPA	EGTA	HEDTA	TTHA
Ag	7.32			6.88	6.71	8.67
Al^{3+}	16.3	17.6	18.6	13.9	14.3	19.7
Ba^{2+}	7.86	8.00	8.87	8.41	6.3	8.22
Be^{2+}	9.30	11.5				
Bi^{3+}	22.8	32.3	35.6		22.3	
Ca^{2+}	10.7	12.1	10.8	10.97	8.3	10.1

Continued table

Metal ion	lg K					
	EDTA	CyDTA	DTPA	EGTA	HEDTA	TTHA
Cd^{2+}	16.7	19.2	19.2	16.7	13.3	19.8
Ce^{3+}	16.0	16.8				
Co^{2+}	16.3	18.9	19.3	12.4	14.6	17.1
Co^{3+}	36.0					
Cr^{3+}	23.4					
Cu^{2+}	18.8	21.3	21.6	17.7	17.6	19.2
Er^{3+}						23.2
Fe^{2+}	14.3	19	16.5	11.9	12.3	
Fe^{3+}	25.1	30.1	28	20.5	19.8	26.8
Ga^{3+}	20.3	22.9	25.5		16.9	
Hg^{2+}	21.8	25.0	26.7	23.2	20.3	26.8
In^{3+}	25.0	28.8	29		20.2	
La^{3+}		16.3				22.2
Li^{+}	2.79					
Mg^{2+}	8.70	11.0	9.30	5.21	7.00	8.43
Mn^{2+}	13.9	16.8	15.6	12.3	10.9	14.7
Mo(Ⅴ)	6.36					
Na^{+}	1.66					
Nd^{3+}	16.6	17.7				22.8
Ni^{2+}	18.6	20.3	20.3	13.6	17.3	18.1
Pb^{2+}	18.0	19.7	18.8	14.7	15.7	17.1
Pd^{2+}	18.5					
Pr^{3+}	16.4	17.3				
Sc^{3+}	23.1	26.1	24.5	18.2		
Sm^{3+}						24.3
Sn^{2+}	18.3					
Sn^{4+}	34.5					
Sr^{2+}	8.73	10.6	9.77	8.5	6.9	9.26
Th^{4+}	23.2	25.6	28.8			31.9

Continued table

Metal ion	lg K					
	EDTA	CyDTA	DTPA	EGTA	HEDTA	TTHA
TiO^{2+}	17.3					
Tl^{3+}	37.8	38.3				
U^{4+}	25.8	27.6	7.69			
VO^{2+}	18.8	19.4				
Y^{3+}	18.1	19.2	22.1	17.2	14.8	
Zn^{2+}	16.5	18.8	18.4	12.7	14.7	
Zr^{2+}	29.5		35.8			
Rare-earth Element	16-20	17-22	19		13-16	

EDTA: ethylenediaminetetra-acetic acid; CyDTA: trans-1, 2-diaminocyclohexanetetraacetic acid; DTPA: diethylenetriaminepentaacetic acid; EGTA: ethyleneglycoldiaminetetra-acetic acid; HEDTA: N-hydroxy-ethylethylene diamine triacetic acid; TTHA: triethylenetetramine hexaacetic acid.

Appendix 8 Standard Electrode Potentials(25 ℃)

Electron pair	Half-reaction	$\varphi^{\ominus}$/V
Li(Ⅰ)—(0)	$Li^+ + e^- \rightleftharpoons Li$	−3.040 1
Cs(Ⅰ)—(0)	$Cs^+ + e^- \rightleftharpoons Cs$	−3.026
Rb(Ⅰ)—(0)	$Rb^+ + e^- \rightleftharpoons Rb$	−2.98
K(Ⅰ)—(0)	$K^+ + e^- \rightleftharpoons K$	−2.931
Ba(Ⅱ)—(0)	$Ba^{2+} + 2e^- \rightleftharpoons Ba$	−2.912
Sr(Ⅱ)—(0)	$Sr^{2+} + 2e^- \rightleftharpoons Sr$	−2.89
Ca(Ⅱ)—(0)	$Ca^{2+} + 2e^- \rightleftharpoons Ca$	−2.868
Na(Ⅰ)—(0)	$Na^+ + e^- \rightleftharpoons Na$	−2.71
La(Ⅲ)—(0)	$La^{3+} + 3e^- \rightleftharpoons La$	−2.379
Mg(Ⅱ)—(0)	$Mg^{2+} + 2e^- \rightleftharpoons Mg$	−2.372
Ce(Ⅲ)—(0)	$Ce^{3+} + 3e^- \rightleftharpoons Ce$	−2.336
H(0)—(−Ⅰ)	$H_2(g) + 2e^- \rightleftharpoons 2H^-$	−2.23
Al(Ⅲ)—(0)	$AlF_6^{3-} + 3e^- \rightleftharpoons Al + 6F^-$	−2.069
Th(Ⅳ)—(0)	$Th^{4+} + 4e^- \rightleftharpoons Th$	−1.899
Be(Ⅱ)—(0)	$Be^{2+} + 2e^- \rightleftharpoons Be$	−1.847

Continued table

Electron pair	Half-reaction	$\varphi^\ominus$/V
U(Ⅲ)—(0)	$U^{3+}+3e^- \rightleftharpoons U$	−1.798
Hf(Ⅳ)—(0)	$HfO^{2+}+2H^++4e^- \rightleftharpoons Hf+H_2O$	−1.724
Al(Ⅲ)—(0)	$Al^{3+}+3e^- \rightleftharpoons Al$	−1.662
Ti(Ⅱ)—(0)	$Ti^{2+}+2e^- \rightleftharpoons Ti$	−1.630
Zr(Ⅳ)—(0)	$ZrO_2+4H^++4e^- \rightleftharpoons Zr+2H_2O$	−1.553
Si(Ⅳ)—(0)	$[SiF_6]^{2-}+4e^- \rightleftharpoons Si+6F^-$	−1.24
Mn(Ⅱ)—(0)	$Mn^{2+}+2e^- \rightleftharpoons Mn$	−1.185
Cr(Ⅱ)—(0)	$Cr^{2+}+2e^- \rightleftharpoons Cr$	−0.913
Ti(Ⅲ)—(Ⅱ)	$Ti^{3+}+e^- \rightleftharpoons Ti^{2+}$	−0.9
B(Ⅲ)—(0)	$H_3BO_3+3H^++3e^- \rightleftharpoons B+3H_2O$	−0.869 8
Ti(Ⅳ)—(0)	$TiO_2+4H^++4e^- \rightleftharpoons Ti+2H_2O$	−0.86
Te(0)—(−Ⅱ)	$Te+2H^++2e^- \rightleftharpoons H_2Te$	−0.793
Zn(Ⅱ)—(0)	$Zn^{2+}+2e^- \rightleftharpoons Zn$	−0.761 8
Ta(Ⅴ)—(0)	$Ta_2O_5+10H^++10e^- \rightleftharpoons 2Ta+5H_2O$	−0.750
Cr(Ⅲ)—(0)	$Cr^{3+}+3e^- \rightleftharpoons Cr$	−0.744
Nb(Ⅴ)—(0)	$Nb_2O_5+10H^++10e^- \rightleftharpoons 2Nb+5H_2O$	−0.644
As(0)—(−Ⅲ)	$As+3H^++3e^- \rightleftharpoons AsH_3$	−0.608
U(Ⅳ)—(Ⅲ)	$U^{4+}+e^- \rightleftharpoons U^{3+}$	−0.607
Ga(Ⅲ)—(0)	$Ga^{3+}+3e^- \rightleftharpoons Ga$	−0.549
P(Ⅰ)—(0)	$H_3PO_2+H^++e^- \rightleftharpoons P+2H_2O$	−0.508
P(Ⅲ)—(Ⅰ)	$H_3PO_3+2H^++2e^- \rightleftharpoons H_3PO_2+H_2O$	−0.499
C(Ⅳ)—(Ⅲ)	$2CO_2+2H^++2e^- \rightleftharpoons H_2C_2O_4$	−0.49
Fe(Ⅱ)—(0)	$Fe^{2+}+2e^- \rightleftharpoons Fe$	−0.447
Cr(Ⅲ)—(Ⅱ)	$Cr^{3+}+e^- \rightleftharpoons Cr^{2+}$	−0.407
Cd(Ⅱ)—(0)	$Cd^{2+}+2e^- \rightleftharpoons Cd$	−0.403 0
Se(0)—(−Ⅱ)	$Se+2H^++2e^- \rightleftharpoons H_2Se(aq)$	−0.399
Pb(Ⅱ)—(0)	$PbI_2+2e^- \rightleftharpoons Pb+2I^-$	−0.365
Eu(Ⅲ)—(Ⅱ)	$Eu^{3+}+e^- \rightleftharpoons Eu^{2+}$	−0.36
Pb(Ⅱ)—(0)	$PbSO_4+2e^- \rightleftharpoons Pb+SO_4^{2-}$	−0.358 8
In(Ⅲ)—(0)	$In^{3+}+3e^- \rightleftharpoons In$	−0.338 2

Continued table

Electron pair	Half-reaction	$\varphi^{\ominus}$/V
Tl(Ⅰ)—(0)	$Tl^+ + e^- \rightleftharpoons Tl$	−0.336
Co(Ⅱ)—(0)	$Co^{2+} + 2e^- \rightleftharpoons Co$	−0.28
P(Ⅴ)—(Ⅲ)	$H_3PO_4 + 2H^+ + 2e^- \rightleftharpoons H_3PO_3 + H_2O$	−0.276
Pb(Ⅱ)—(0)	$PbCl_2 + 2e^- \rightleftharpoons Pb + 2Cl^-$	−0.267 5
Ni (Ⅱ)—(0)	$Ni^{2+} + 2e^- \rightleftharpoons Ni$	−0.257
V(Ⅲ)—(Ⅱ)	$V^{3+} + e^- \rightleftharpoons V^{2+}$	−0.255
Ge(Ⅳ)—(0)	$H_2GeO_3 + 4H^+ + 4e^- \rightleftharpoons Ge + 3H_2O$	−0.182
Ag(Ⅰ)—(0)	$AgI + e^- \rightleftharpoons Ag + I^-$	−0.152 24
Sn(Ⅱ)—(0)	$Sn^{2+} + 2e^- \rightleftharpoons Sn$	−0.137 5
Pb(Ⅱ)—(0)	$Pb^{2+} + 2e^- \rightleftharpoons Pb$	−0.126 2
C(Ⅳ)—(Ⅱ)	$CO_2(g) + 2H^+ + 2e^- \rightleftharpoons CO + H_2O$	−0.12
P(0)—(−Ⅲ)	$P(\text{white}) + 3H^+ + 3e^- \rightleftharpoons PH_3(g)$	−0.063
Hg(Ⅰ)—(0)	$Hg_2I_2 + 2e^- \rightleftharpoons 2Hg + 2I^-$	−0.040 5
Fe(Ⅲ)—(0)	$Fe^{3+} + 3e^- \rightleftharpoons Fe$	−0.037
H(Ⅰ)—(0)	$2H^+ + 2e^- \rightleftharpoons H_2$	0.000 0
Ag(Ⅰ)—(0)	$AgBr + e^- \rightleftharpoons Ag + Br^-$	0.071 33
S(Ⅱ.Ⅴ)—(Ⅱ)	$S_4O_6^{2-} + 2e^- \rightleftharpoons 2S_2O_3^{2-}$	0.08
Ti(Ⅳ)—(Ⅲ)	$TiO^{2+} + 2H^+ + e^- \rightleftharpoons Ti^{3+} + H_2O$	0.1
S(0)—(−Ⅱ)	$S + 2H^+ + 2e^- \rightleftharpoons H_2S(aq)$	0.142
Sn(Ⅳ)—(Ⅱ)	$Sn^{4+} + 2e^- \rightleftharpoons Sn^{2+}$	0.151
Sb(Ⅲ)—(0)	$Sb_2O_3 + 6H^+ + 6e^- \rightleftharpoons 2Sb + 3H_2O$	0.152
Bi(Ⅲ)—(0)	$BiOCl + 2H^+ + 3e^- \rightleftharpoons Bi + Cl^- + H_2O$	0.158 3
S(Ⅵ)—(Ⅳ)	$SO_4^{2-} + 4H^+ + 2e^- \rightleftharpoons H_2SO_3 + H_2O$	0.172
Sb(Ⅲ)—(0)	$SbO^+ + 2H^+ + 3e^- \rightleftharpoons Sb + H_2O$	0.212
Ag(Ⅰ)—(0)	$AgCl + e^- \rightleftharpoons Ag + Cl^-$	0.222 33
As(Ⅲ)—(0)	$HAsO_2 + 3H^+ + 3e^- \rightleftharpoons As + 2H_2O$	0.248
Hg(Ⅰ)—(0)	$Hg_2Cl_2 + 2e^- \rightleftharpoons 2Hg + 2Cl^-$ (saturation KCl)	0.268 08
Bi(Ⅲ)—(0)	$BiO^+ + 2H^+ + 3e^- \rightleftharpoons Bi + H_2O$	0.320
U(Ⅵ)—(Ⅳ)	$UO_2^{2+} + 4H^+ + 2e^- \rightleftharpoons U^{4+} + 2H_2O$	0.327
C(Ⅳ)—(Ⅲ)	$2HCNO + 2H^+ + 2e^- \rightleftharpoons (CN)_2 + 2H_2O$	0.330

Continued table

Electron pair	Half-reaction	$\varphi^{\ominus}$/V
V(Ⅳ)—(Ⅲ)	$VO^{2+}+2H^{+}+e^{-} \rightleftharpoons V^{3+}+H_2O$	0.337
Cu(Ⅱ)—(0)	$Cu^{2+}+2e^{-} \rightleftharpoons Cu$	0.341 9
Re(Ⅶ)—(0)	$ReO_4^{-}+8H^{+}+7e^{-} \rightleftharpoons Re+4H_2O$	0.368
Ag(Ⅰ)—(0)	$Ag_2CrO_4+2e^{-} \rightleftharpoons 2Ag+CrO_4^{2-}$	0.447 0
S(Ⅳ)—(0)	$H_2SO_3+4H^{+}+4e^{-} \rightleftharpoons S+3H_2O$	0.449
Cu(Ⅰ)—(0)	$Cu^{+}+e^{-} \rightleftharpoons Cu$	0.521
I(0)—(−Ⅰ)	$I_2+2e^{-} \rightleftharpoons 2I^{-}$	0.535 5
I(0)—(−Ⅰ)	$I_3^{-}+2e^{-} \rightleftharpoons 3I^{-}$	0.536
As(Ⅴ)—(Ⅲ)	$H_3AsO_4+2H^{+}+2e^{-} \rightleftharpoons HAsO_2+2H_2O$	0.560
Sb(Ⅴ)—(Ⅲ)	$Sb_2O_5+6H^{+}+4e^{-} \rightleftharpoons 2SbO^{+}+3H_2O$	0.581
Te(Ⅳ)—(0)	$TeO_2+4H^{+}+4e^{-} \rightleftharpoons Te+2H_2O$	0.593
U(Ⅴ)—(Ⅳ)	$UO_2^{+}+4H^{+}+e^{-} \rightleftharpoons U^{4+}+2H_2O$	0.612
Hg(Ⅱ)—(Ⅰ)	$2HgCl_2+2e^{-} \rightleftharpoons Hg_2Cl_2+2Cl^{-}$	0.63
Pt(Ⅳ)—(Ⅱ)	$[PtCl_6]^{2-}+2e^{-} \rightleftharpoons [PtCl_4]^{2-}+2Cl^{-}$	0.68
O(0)—(−Ⅰ)	$O_2+2H^{+}+2e^{-} \rightleftharpoons H_2O_2$	0.695
Pt(Ⅱ)—(0)	$[PtCl_4]^{2-}+2e^{-} \rightleftharpoons Pt+4Cl^{-}$	0.755
Se(Ⅳ)—(0)	$H_2SeO_3+4H^{+}+4e^{-} \rightleftharpoons Se+3H_2O$	0.74
Fe(Ⅲ)—(Ⅱ)	$Fe^{3+}+e^{-} \rightleftharpoons Fe^{2+}$	0.771
Hg(Ⅰ)—(0)	$Hg_2^{2+}+2e^{-} \rightleftharpoons 2Hg$	0.797 3
Ag(Ⅰ)—(0)	$Ag^{+}+e^{-} \rightleftharpoons Ag$	0.799 6
Os(Ⅷ)—(0)	$OsO_4+8H^{+}+8e^{-} \rightleftharpoons Os+4H_2O$	0.8
N(Ⅴ)—(Ⅳ)	$2NO_3^{-}+4H^{+}+2e^{-} \rightleftharpoons N_2O_4+2H_2O$	0.803
Hg(Ⅱ)—(0)	$Hg^{2+}+2e^{-} \rightleftharpoons Hg$	0.851
Si(Ⅳ)—(0)	(quartz)$SiO_2+4H^{+}+4e^{-} \rightleftharpoons Si+2H_2O$	0.857
Cu(Ⅱ)—(Ⅰ)	$Cu^{2+}+I^{-}+e^{-} \rightleftharpoons CuI$	0.86
N(Ⅲ)—(Ⅰ)	$2HNO_2+4H^{+}+4e^{-} \rightleftharpoons H_2N_2O_2+2H_2O$	0.86
N(Ⅴ)—(Ⅲ)	$NO_3^{-}+3H^{+}+2e^{-} \rightleftharpoons HNO_2+H_2O$	0.934
Pd(Ⅱ)—(0)	$Pd^{2+}+2e^{-} \rightleftharpoons Pd$	0.951
N(Ⅴ)—(Ⅱ)	$NO_3^{-}+4H^{+}+3e^{-} \rightleftharpoons NO+2H_2O$	0.957

Continued table

Electron pair	Half-reaction	$\varphi^{\ominus}/V$
N(Ⅲ)—(Ⅱ)	$HNO_2+H^++e^- \rightleftharpoons NO+H_2O$	0.983
I(Ⅰ)—(−Ⅰ)	$HIO+H^++2e^- \rightleftharpoons I^-+H_2O$	0.987
V(Ⅴ)—(Ⅳ)	$VO_2^++2H^++e^- \rightleftharpoons VO^{2+}+H_2O$	0.991
V(Ⅴ)—(Ⅳ)	$V(OH)_4^++2H^++e^- \rightleftharpoons VO^{2+}+3H_2O$	1.00
Au(Ⅲ)—(0)	$[AuCl_4]^-+3e^- \rightleftharpoons Au+4Cl^-$	1.002
Te(Ⅵ)—(Ⅳ)	$H_6TeO_6+2H^++2e^- \rightleftharpoons TeO_2+4H_2O$	1.02
N(Ⅳ)—(Ⅱ)	$N_2O_4+4H^++4e^- \rightleftharpoons 2NO+2H_2O$	1.035
N(Ⅳ)—(Ⅲ)	$N_2O_4+2H^++2e^- \rightleftharpoons 2HNO_2$	1.065
I(Ⅴ)—(−Ⅰ)	$IO_3^-+6H^++6e^- \rightleftharpoons I^-+3H_2O$	1.085
Br(0)—(−Ⅰ)	$Br_2(aq)+2e^- \rightleftharpoons 2Br^-$	1.087 3
Se(Ⅵ)—(Ⅳ)	$SeO_4^{2-}+4H^++2e^- \rightleftharpoons H_2SeO_3+H_2O$	1.151
Cl(Ⅴ)—(Ⅳ)	$ClO_3^-+2H^++e^- \rightleftharpoons ClO_2+H_2O$	1.152
Pt(Ⅱ)—(0)	$Pt^{2+}+2e^- \rightleftharpoons Pt$	1.18
Cl(Ⅶ)—(Ⅴ)	$ClO_4^-+2H^++2e^- \rightleftharpoons ClO_3^-+H_2O$	1.189
I(Ⅴ)—(0)	$2IO_3^-+12H^++10e^- \rightleftharpoons I_2+6H_2O$	1.195
Cl(Ⅴ)—(Ⅲ)	$ClO_3^-+3H^++2e^- \rightleftharpoons HClO_2+H_2O$	1.214
Mn(Ⅳ)—(Ⅱ)	$MnO_2+4H^++2e^- \rightleftharpoons Mn^{2+}+2H_2O$	1.224
O(0)—(−Ⅱ)	$O_2+4H^++4e^- \rightleftharpoons 2H_2O$	1.229
Tl(Ⅲ)—(Ⅰ)	$Tl^{3+}+2e^- \rightleftharpoons Tl^+$	1.252
Cl(Ⅳ)—(Ⅲ)	$ClO_2+H^++e^- \rightleftharpoons HClO_2$	1.277
N(Ⅲ)—(Ⅰ)	$2HNO_2+4H^++4e^- \rightleftharpoons N_2O+3H_2O$	1.297
Cr(Ⅵ)—(Ⅲ)	$Cr_2O_7^{2-}+14H^++6e^- \rightleftharpoons 2Cr^{3+}+7H_2O$	1.33
Br(Ⅰ)—(−Ⅰ)	$HBrO+H^++2e^- \rightleftharpoons Br^-+H_2O$	1.331
Cr(Ⅵ)—(Ⅲ)	$HCrO_4^-+7H^++3e^- \rightleftharpoons Cr^{3+}+4H_2O$	1.350
Cl(0)—(−Ⅰ)	$Cl_2(g)+2e^- \rightleftharpoons 2Cl^-$	1.358 27
Cl(Ⅶ)—(−Ⅰ)	$ClO_4^-+8H^++8e^- \rightleftharpoons Cl^-+4H_2O$	1.389
Cl(Ⅶ)—(0)	$ClO_4^-+8H^++7e^- \rightleftharpoons 1/2Cl_2+4H_2O$	1.39
Au(Ⅲ)—(Ⅰ)	$Au^{3+}+2e^- \rightleftharpoons Au^+$	1.401
Br(Ⅴ)—(−Ⅰ)	$BrO_3^-+6H^++6e^- \rightleftharpoons Br^-+3H_2O$	1.423
I(Ⅰ)—(0)	$2HIO+2H^++2e^- \rightleftharpoons I_2+2H_2O$	1.439
Cl(Ⅴ)—(−Ⅰ)	$ClO_3^-+6H^++6e^- \rightleftharpoons Cl^-+3H_2O$	1.451
Pb(Ⅳ)—(Ⅱ)	$PbO_2+4H^++2e^- \rightleftharpoons Pb^{2+}+2H_2O$	1.455

Continued table

Electron pair	Half-reaction	$\varphi^{\ominus}$/V
Cl(Ⅰ)—(−Ⅰ)	$HClO+H^{+}+2e^{-} \rightleftharpoons Cl^{-}+H_2O$	1.482
Br(Ⅴ)—(0)	$BrO_3^{-}+6H^{+}+5e^{-} \rightleftharpoons 1/2Br_2+3H_2O$	1.482
Au(Ⅲ)—(0)	$Au^{3+}+3e^{-} \rightleftharpoons Au$	1.498
Mn(Ⅶ)—(Ⅱ)	$MnO_4^{-}+8H^{+}+5e^{-} \rightleftharpoons Mn^{2+}+4H_2O$	1.507
Mn(Ⅲ)—(Ⅱ)	$Mn^{3+}+e^{-} \rightleftharpoons Mn^{2+}$	1.5415
Cl(Ⅲ)—(−Ⅰ)	$HClO_2+3H^{+}+4e^{-} \rightleftharpoons Cl^{-}+2H_2O$	1.570
Br(Ⅰ)—(0)	$HBrO+H^{+}+e^{-} \rightleftharpoons 1/2Br_2(aq)+H_2O$	1.574
N(Ⅱ)—(Ⅰ)	$2NO+2H^{+}+2e^{-} \rightleftharpoons N_2O+H_2O$	1.591
I(Ⅶ)—(Ⅴ)	$H_5IO_6+H^{+}+2e^{-} \rightleftharpoons IO_3^{-}+3H_2O$	1.601
Cl(Ⅰ)—(0)	$HClO+H^{+}+e^{-} \rightleftharpoons 1/2Cl_2+H_2O$	1.611
Cl(Ⅲ)—(Ⅰ)	$HClO_2+2H^{+}+2e^{-} \rightleftharpoons HClO+H_2O$	1.645
Ni(Ⅳ)—(Ⅱ)	$NiO_2+4H^{+}+2e^{-} \rightleftharpoons Ni^{2+}+2H_2O$	1.678
Mn(Ⅶ)—(Ⅳ)	$MnO_4^{-}+4H^{+}+3e^{-} \rightleftharpoons MnO_2+2H_2O$	1.679
Pb(Ⅳ)—(Ⅱ)	$PbO_2+SO_4^{2-}+4H^{+}+2e^{-} \rightleftharpoons PbSO_4+2H_2O$	1.691 3
Au(Ⅰ)—(0)	$Au^{+}+e^{-} \rightleftharpoons Au$	1.692
Ce(Ⅳ)—(Ⅲ)	$Ce^{4+}+e^{-} \rightleftharpoons Ce^{3+}$	1.72
N(Ⅰ)—(0)	$N_2O+2H^{+}+2e^{-} \rightleftharpoons N_2+H_2O$	1.766
O(-Ⅰ)—(−Ⅱ)	$H_2O_2+2H^{+}+2e^{-} \rightleftharpoons 2H_2O$	1.776
Co(Ⅲ)—(Ⅱ)	$Co^{3+}+e^{-} \rightleftharpoons Co^{2+}$ (2 mol·L^{-1} H_2SO_4)	1.83
Ag(Ⅱ)—(Ⅰ)	$Ag^{2+}+e^{-} \rightleftharpoons Ag^{+}$	1.980
S(Ⅶ)—(Ⅵ)	$S_2O_8^{2-}+2e^{-} \rightleftharpoons 2SO_4^{2-}$	2.010
O(0)—(−Ⅱ)	$O_3+2H^{+}+2e^{-} \rightleftharpoons O_2+H_2O$	2.076
O(Ⅱ)—(−Ⅱ)	$F_2O+2H^{+}+4e^{-} \rightleftharpoons H_2O+2F^{-}$	2.153
Fe(Ⅵ)—(Ⅲ)	$FeO_4^{2-}+8H^{+}+3e^{-} \rightleftharpoons Fe^{3+}+4H_2O$	2.20
O(0)—(−Ⅱ)	$O(g)+2H^{+}+2e^{-} \rightleftharpoons H_2O$	2.421
F(0)—(−Ⅰ)	$F_2+2e^{-} \rightleftharpoons 2F^{-}$	2.866
	$F_2+2H^{+}+2e^{-} \rightleftharpoons 2HF$	3.053
Cu(Ⅱ)—(Ⅰ)	$Cu^{2+}+e^{-} \rightleftharpoons Cu^{+}$	0.17

Appendix 9 Formal Potentials of Some Oxidation-Reduction Electron Pairs

Half-reaction	Formal potential/V	Buffer/($mol \cdot L^{-1}$)
$Ag(II) + e^- \rightleftharpoons Ag^+$	1.927	$c(HNO_3)=4$
	2.00	$c(HClO_4)=4$
$Ag^+ + e^- \rightleftharpoons Ag$	0.792	$c(HClO_4)=1$
	0.228	$c(HCl)=1$
	0.59	$c(NaOH)=1$
$Ce^{4+} + e^- \rightleftharpoons Ce^{3+}$	1.70	$c(HClO_4)=1$
	1.61	$c(HNO_3)=1$
	1.44	$c(H_2SO_4)=0.5$
	1.28	$c(HCl)=1$
$Co^{3+} + e^- \rightleftharpoons Co^{2+}$	1.84	$c(HNO_3)=3$
$Cr^{3+} + e^- \rightleftharpoons Cr^{2+}$	−0.40	$c(HCl)=5$
$Cr_2O_7^{2-} + 14H^+ + 6e^- \rightleftharpoons 2Cr^{3+} + 7H_2O$	0.93	$c(HCl)=0.1$
	0.97	$c(HCl)=0.5$
	1.0	$c(HCl)=1$
	1.05	$c(HCl)=2$
	1.08	$c(HCl)=3$
	1.11	$c(H_2SO_4)=2$
	1.15	$c(H_2SO_4)=4$
	1.30	$c(H_2SO_4)=6$
	1.34	$c(H_2SO_4)=8$
	0.84	$c(HClO_4)=0.1$
	1.025	$c(HClO_4)=1$
	1.27	$c(HNO_3)=0.1$
$CrO_4^{2-} + 2H_2O + 3e^- \rightleftharpoons CrO_2^- + 4OH^-$	−0.12	$c(NaOH)=1$
$Cu^{2+} + e^- \rightleftharpoons Cu^+$	−0.09	pH=14

Continued table

Half-reaction	Formal potential/V	Buffer/$(mol \cdot L^{-1})$
$Fe^{3+} + e^- \rightleftharpoons Fe^{2+}$	0.75	$c(HClO_4)=1$
	0.68	$c(H_2SO_4)=1$
	0.70	$c(HCl)=1$
	0.46	$c(H_3PO_4)=2$
	0.51	$c(HCl)=1$
		$c(H_3PO_4)=0.25$
$H_3AsO_4 + 2H^+ + 2e^- \rightleftharpoons H_3AsO_3 + 2H_2O$	0.557	$c(HCl)=1$
	0.557	$c(HClO_4)=1$
$MnO_4^- + 8H^+ + 5e^- \rightleftharpoons Mn^{2+} + 4H_2O$	1.45	$c(HClO_4)=1$
	1.27	$c(H_3PO_4)=8$
$SnCl_6^{2-} + 2e^- \rightleftharpoons SnCl_4^{2-} + 2Cl^-$	0.14	$c(HCl)=1$
$Sn^{2+} + 2e^- \rightleftharpoons Sn$	−0.16	$c(HClO_4)=1$
$Pb^{2+} + 2e^- \rightleftharpoons Pb$	−0.32	$c(NaAc)=1$
	−0.14	$c(HClO_4)=1$
$Ti(IV) + e^- \rightleftharpoons Ti(III)$	−0.01	$c(H_2SO_4)=0.4$
	0.12	$c(H_2SO_4)=4$
	−0.04	$c(HCl)=1$
	−0.05	$c(H_3PO_4)=1$

References

[1] Hage D S, Carr J D. Analytical Chemical and Quantitative Analysis [M]. London: Pearson, 2010.

[2] Skoog D A, West D M, Holler F J, et al. Fundamentals of Analytical Chemistry [M]. 9th ed. Boston: Cengage Learning, 2013.

[3] Christian G D, Dasgupta P K, Schug K A. Analytical Chemistry [M]. 7th ed. John Wiley & Sons, Inc, 2013.

[4] Harris D C. Analytical Chemistry and Quantitative Chemistry [M]. 10th ed. San Francisco: W. H. Freeman, 2019.

[5] Hu P, Wang Q. Instrumental Analysis [M]. 5th ed. Beijing: Higher Education Press, 2019. (in Chinese)

[6] Patnaik P. Dean's Analytical Chemistry Handbook [M]. 2nd ed. New York: McGraw-Hill, 2004.

[7] Qin X. General Chemistry [M]. Beijing: China Agriculture Press, 2021. (in Chinese)

[8] Zhou H, Qin X, Li X, et al. Guidelines and Exercise Answers for General Chemistry and Quantitative Analytical Chemistry [M]. Beijing: China Agriculture Press, 2022. (in Chinese)

[9] Zhou H. Quantitative Analytical Chemistry [M]. Beijing: China Agriculture Press, 2020. (in Chinese)